MOTHMAN

TO WRITE TO THE AUTHORS

If you wish to contact the author or would like more information about this book, please write to the author in care of Llewellyn Worldwide Ltd. and we will forward your request. Both the author and the publisher appreciate hearing from you and learning of your enjoyment of this book and how it has helped you. Llewellyn Worldwide Ltd. cannot guarantee that every letter written to the author can be answered, but all will be forwarded. Please write to:

Richard Estep and/or Tobias Wayland
℅ Llewellyn Worldwide
2143 Wooddale Drive
Woodbury, MN 55125-2989

Please enclose a self-addressed stamped envelope for reply, or $1.00 to cover costs. If outside the U.S.A., enclose an international postal reply coupon.

Many of Llewellyn's authors have websites with additional information and resources. For more information, please visit our website at http://www.llewellyn.com.

PRAISE FOR *MOTHMAN*

"Richard and Tobias didn't just write a book. They tapped into something that's been lurking in the shadows of our reality for way too long. As someone who's had unexplainable experiences myself, this one felt incredibly personal. It's both captivatingly weird and unsettling, and it makes you feel a little less crazy for wondering if it all really happened."

—**SHANE PITTMAN**, author and paranormal investigator and cast member of *The Holzer Files* and *28 Days Haunted*

"Paranormal pros Richard Estep and Tobias Wayland chase the truth behind America's most chilling cryptid. From eerie sightings in Point Pleasant to a new wave of terror near Chicago, this book uncovers shocking encounters and haunting clues that keep the legend alive. A must read for anyone fascinated by the legend, the myths, and the reality behind this enigma."

—**DAVE SCHRADER**, host of the *Paranormal 360* radio show

MOTHMAN

SIGHTINGS AND INVESTIGATIONS OF THE ICONIC FLYING CRYPTID

RICHARD ESTEP TOBIAS WAYLAND

WOODBURY, MINNESOTA

First Edition
First Printing, 2025

Book design by Rordan Brasington
Cover art by Jan Meininghaus
Cover design by Kevin R. Brown
Interior illustrations by Llewellyn Art Department

Library of Congress Cataloging-in-Publication Data (Pending)
ISBN: 978-0-7387-7990-4

Llewellyn Publications
A Division of Llewellyn Worldwide Ltd.
2143 Wooddale Drive
Woodbury, MN 55125-2989
www.llewellyn.com

Printed in the United States of America

GPSR Representation:
UPI-2M PLUS d.o.o., Medulićeva 20, 10000 Zagreb, Croatia
matt.parsons@upi2mbooks.hr

OTHER BOOKS BY RICHARD ESTEP

In Search of Demons

The Horrors of Fox Hollow Farm

OTHER BOOKS BY TOBIAS WAYLAND

The Lake Michigan Mothman:
High Strangeness in the Midwest
(Singular Fortean Publishing, 2019)

Strange Tales of the Impossible
(Singular Fortean Publishing, 2021)

The Singular Fortean Society's
Yuletide Guide to High Strangeness
(Singular Fortean Publishing, 2023)

© AliCottonphoto.com

ABOUT THE AUTHOR

Richard Estep has been a paranormal investigator for thirty years, researching claims of ghosts and hauntings on both sides of the Atlantic. He is the author of more than thirty books, in genres ranging from paranormal nonfiction and true crime to history and fiction. He is a cast member of the TV shows *Haunted Hospitals*, *Paranormal Revenge*, and *Paranormal 911*, and has made guest appearances on *A Haunting*, *Destination Fear*, and *Haunted Case Files*.

Richard writes a regular column for *Haunted Magazine* and frequently presents at conferences and public events across the United States. A paramedic by profession, Richard lives in Colorado with his family. You can visit him at RichardEstep.net.

© Emily Wayland

ABOUT THE AUTHOR

Tobias Wayland is a passionate Fortean who has been actively investigating the unusual for over a decade. The first several years of his investigative career were spent as a MUFON field investigator, and following that he investigated independently prior to becoming the head writer and editor for The Singular Fortean Society. Tobias is a frequent guest on various podcasts and radio shows, has written several books and contributed articles to periodicals on the paranormal, has appeared on television and in documentaries, and is often invited to speak at paranormal conferences and events.

He was featured in season four of *Unsolved Mysteries*, the series premiere of *Expedition X*, and the Small Town

Monsters documentaries *Terror in the Skies* and *On the Trail of the Lake Michigan Mothman* for his work investigating Mothman sightings in the Midwest. He and his wife, Emily, have been involved with the Lake Michigan Mothman investigation since its advent in the spring of 2017 and published a book chronicling the experience, *The Lake Michigan Mothman: High Strangeness in the Midwest*. His second and third books about unusual phenomena, *Strange Tales of the Impossible* and *The Singular Fortean Society's Yuletide Guide to High Strangeness*, continued their work in investigating a variety of seemingly impossible events.

His years as an investigator have served him best by illustrating that when it comes to the anomalous, the preternatural, and the paranormal, any answers he's found are still hopelessly outnumbered by questions.

Dedicated to the memory of David Taylor
Gone too soon, but never forgotten

CONTENTS

FOREWORD

On a chilly autumn evening in 2024, I sat down to speak with a gentleman named Bob. Through a mutual, trusted connection, Bob had agreed to tell me about his encounter with a strange flying creature. It was a story he had rarely shared and never with someone outside of his family.

Bob was a clean-cut man in his sixties who had built his own business, raised a family, and lived a typical blue-collar life. He was clearly hesitant to share his account, but at the same time, he said he was glad to finally tell someone else about what he'd seen that fall day.

In 1986, Bob had traveled from his home in Pennsylvania down to West Virginia to do some hunting. He stayed the night with a friend, and the next morning, they were up before dawn to see if they could bag some deer. Just after sunrise, they found themselves moving through a wooded area. About an hour into their time in the woods, Bob had an experience that has stayed with him his entire life.

Bob took a slow sip of his coffee and began to relate his story. He was direct and straightforward and quickly got to the encounter itself. He recalled:

"We were out hunting. Honestly, we were hunting in a place we should not have been—legally that is. My hunting buddy was off to the west, and at the time he was out of sight, but we were walking parallel, and we were frequent hunting partners, so we felt safe that day even though we weren't wearing orange jackets.

"I was in a very small clearing and about to head into some more trees when I heard something rustling ahead. I stopped and stood completely still. Whatever it was, it sounded large, so I thought for sure it was a deer. In the next minute, I saw something dark moving around, but it looked like it was moving up from the ground which was odd.

"This thing went straight up, and it landed on the branch of a big tree directly in front of me. It was in an open space that made a perfect perch for it. At first, I thought maybe it was a really, really big owl, but I quickly realized that it was too big for that. This thing was man sized and partly man shaped with two long legs and a pair of human-type arms, and it had wings. In fact, it had a massive wing-span and when it landed, it folded its wings in part way. You could see them sticking out the sides and up over the shoulders. It had kind of a menacing hunch to it. The head—it was odd. I mean, there was a head there, but it sort of sat too far down on the shoulders like it was somehow partly tucked into the upper body. Maybe it was the wings that caused it to look like that—I don't know.

"I raised my gun up and I wanted to shoot it, I really did, but the only thing I could think about was what if I missed? I held the gun for a minute then just as I started to lower it, that thing lifted up and took to the air, and it let out a screech. It was something like a cross between a bird of prey and a woman screaming.

"It flew—and I ran."

Bob told me that he looked back long enough to see the creature flying in the opposite direction, but that didn't prevent him from running all the way back to the truck and climbing into the cab, where he remained until his friend returned. When the other man came back, Bob just said that he'd gotten sick and had to rest. He never told his hunting partner about the creature he'd witnessed.

Bob did, eventually, tell his wife and another family member, but beyond this, he never discussed his experience. I pressed him for more details, and as we talked, he brought up the Mothman.

"I never, never believed any of that nonsense about the Mothman," he said. "I'd heard the stories back in the day but thought they were ridiculous. I always believed there were two kinds of animals—domestic animals that you raised or kept as pets—and wild animals that could be hunted and shot, but that thing, that thing wasn't either of those. The creature I saw that day was not of this earth and I pray to God that I never see it again."

The Mothman "nonsense" that Bob refers to is the very center piece of the book that you're about to delve into. Bob's tale is one of countless accounts involving a strange, winged creature that has become a true legend, and his report gives us an insight into the very nature of the Mothman's effect on those who witness it. Widely considered a "cryptid" or undiscovered animal by many people, the creature is one that truly skirts the edges of various types of phenomena, making it a true puzzle even to those who occupy themselves with the study of strange things.

The origins of the creature as we know them date back to 1966 when residents of the small town of Point Pleasant, West Virginia (and some neighbors across the river in Ohio) began to report a bizarre, winged monster that was terrorizing locals. The monster's presence in the town culminated with the collapse of the Silver Bridge in December 1967.

The Mothman was brought to widespread attention due to two things—a book titled *The Mothman Prophecies,* published in 1975, and the subsequent film of the same name from 2002. The book that started it all was written by John Keel, a well-known Fortean writer who spent time in Point Pleasant investigating the sightings.

Keel is considered a controversial figure, as you'll see reflected in this book. Loved by many, disparaged by some, he remains a prominent individual in the study of high strangeness. I'll admit a certain bias myself. I knew John personally and consider him both an important figure in terms of his contributions to the Fortean field and an influence on my own writing.

In the decades since Keel wrote about the Mothman, the creature has turned up on television, in documentaries, and in countless publications. There's a good reason for this. The creature is not only fascinating, but it seems that it has lingered in other ways as well. Over the years, people have reported encounters with Mothman-like creatures all over the world. Whether or not this is the original beast, its offspring, or something else entirely is anyone's guess, and just like with the original encounters, the mysteries are not simple ones, and often, other strange effects come hand in hand with the winged thing.

Physical creature, omen, harbinger of doom, flight of fancy, real or tangible, these and other possibilities float around the winged beast of Point Pleasant, and with each new encounter, we're left to wonder again exactly what we're dealing with. The cases and the potentials seem endless.

Fortunately, my colleagues Tobias Wayland and Richard Estep have jumped down the rabbit hole that is the Mothman phenomenon. They've traveled, they've talked to people, they've dug through archives, and they've produced an overview that provides a valuable addition to the lore of the creature and its many offshoots.

Whether you're a newcomer to the study of such things or a seasoned researcher, you'll find something to spark your mind and your wonder in *Mothman*. Enjoy the flight!

David Weatherly
Eerielights.com

INTRODUCTION

Mid-century America was full of monsters, and among them, none have persevered like Mothman. To exist alongside such oddities as the Loveland Frogman, Flatwoods Monster, and Kelly-Hopkinsville Goblins and make it out of the twentieth century as one of Forteana's most recognizable monsters is no mean feat. It takes something special, something weird. But it also takes a little help.

There's a running joke among Forteans that John Keel's *The Mothman Prophecies* would barely be considered a pamphlet, let alone a book, if all it contained were actual reports of the Mothman. It's everything else included by Keel in that work that makes it so captivating. That's the allure of the Mothman: not necessarily what it is, but what it represents, and what that empowers us to imagine as possible.

Keel wove a tale of beleaguered investigators harassed by entities beyond the ken of humanity, embattled townsfolk in fear of monsters, and a small town adrift amidst a sea of high strangeness. He combined UFO sightings (and even their purported occupants), other strange creature sightings, psychic and parapsychological phenomena, and assorted folklore with his own Ultraterrestrial hypothesis (which asserts that many unexplained phenomena represent an ancient hidden race of beings with whom we have shared our planet for millennia) into a fantastical narrative—one so immersive and engaging that it has left readers enchanted for decades since. And yet the story remains relatable.

Perhaps most importantly, Mothman takes the impossible and grounds it in the everyday. The witnesses to this phenomenon could be anyone. They could work in your office, go to your church, attend your school, or pass you on the street and you wouldn't look twice. These witnesses are noteworthy because they are us. An encounter with Mothman could happen to anyone and often occurs when we least expect it. This phenomenon is far from over and Mothman sightings continue across the country into the present day. From coast to coast and throughout the Great Lakes, winged humanoid sightings are reported by everyday people just trying to live their lives.

These encounters aren't sought out; they come as we go about our business. Imagine running errands only to find yourself fully immersed in the high strangeness of a winged humanoid encounter through no fault of your own. You've seen something that has shaken you to your core, fully unexpected but no less impactful for it, and now your mind is reeling with no idea how to proceed. You want to share this experience with someone—we're social animals, after all, and the catharsis of sharing our experiences cannot be understated—but you're afraid. What if you're mocked? Could you lose friends? What about your job?

Steeling yourself, you approach a loved one and bare your soul. You desperately want, no, *need* to be taken seriously by this person you would otherwise trust with anything, but instead, you're laughed at. Maybe they don't exactly ostracize you, but their flippant attitude displayed in the face of your vulnerability hurts all the same. If you can't share your experience with this person, then you can't share it with anyone. So, you shut up about it. You keep this experience—this powerful, maybe even traumatic, experience—to yourself and you don't talk about it. Until, one day, you read an article or hear a story or see a television show about someone who has had a similar experience—something you never imagined possible. You thought you were all alone.

Mothman is somewhat known among weirdos, although still nowhere near the level of ghosts or Bigfoot, but you're not a weirdo. You'd never heard of it. You never imagined that anyone had ever seen something similar. But here you are, learning that someone has, and just as importantly, learning that there are people out there who will take you seriously.

So, you reach out to one of the investigators involved in whatever case you've found, and you speak to them. And finally, you achieve catharsis. After who knows how many weeks or months or years, you're able to dump the psychic load you've been carrying on someone actually prepared to hear it. You feel better, and your experience is added to the canon of events to help understand what exactly is going on.

That's why this is important, and it's why we do what we do. If you're reading this, we place you firmly within our camp: the investigators, the researchers; people who have either experienced the impossible for themselves or are at least curious and caring enough to explore these possibilities in the name of humanity. Welcome. Join us. We are about to recount our journey into the unknown.

Truly,
Tobias Wayland and Richard Estep
January 6, 2025

CHAPTER 1

FORTEANA 101

There can be no understanding John Keel without first understanding Charles Fort. And not just Keel, either. Charles Fort is to anomalous studies what H. P. Lovecraft is to horror; that is to say, arguably the most influential person in his field whom most people have never heard of.

Fort was an early twentieth-century collector of weird—but ostensibly true—stories. He spent endless hours in the New York Public Library and British Museum Library researching accounts of the anomalous, finding them most often in scientific journals and newspaper articles. These accounts were compiled into volumes in which Fort would look for commonalities and speculate about what might be behind it all.

His first book on weird phenomena was published in 1919 at the age of forty-five. Entitled *The Book of the Damned*, it covered a variety of strange objects that seemingly appeared from the sky and first displayed one of the hallmarks of Fort's writing—a rebellious refusal to acknowledge the authority of science and religion.

"A procession of the damned," he began. "By the damned, I mean the excluded. We shall have a procession of data that Science has excluded. Battalions of the accursed, captained by pallid data that I have exhumed, will march. You'll read them—or they'll march. Some of them livid and some of them fiery and some of them rotten."[1]

Fort was incensed by what he viewed as the dismissive attitudes displayed by these institutions. According to Fort, the main motivations of religion and science are not discovery or the betterment of humanity (be that spiritual or temporal), but the maintenance of their supremacy in the social hierarchy. To Fort, religion and science are two sides of the same coin, both dogmatically refusing to acknowledge anything inconvenient to their narrative.

As he wrote in his 1931 work *Lo!*, "Witchcraft always has a hard time, until it becomes established and changes its name. We hear much of the conflict between science and religion, but our conflict is with both of these. Science and religion always have agreed in opposing and suppress-

---

1. Fort, *Complete Books of Charles Fort*, 12.

ing the various witchcrafts. Now that religion is inglorious, one of the most fantastic of transferences of worships is that of glorifying science, as a beneficent being. It is the attributing of all that is of development, or of possible betterment to science. But no scientist has ever upheld a new idea, without bringing upon himself abuse from other scientists. Science has done its utmost to prevent whatever science has done."[2]

This general distrust of established authorities is still widespread in anomalous studies today, although it would be difficult to attribute that solely to Fort's influence when the dismissal of mainstream attitudes toward these phenomena is essential to take them seriously enough to study in the first place.

However, other of his contributions are far more easily attributable.

For instance, Fort coined the term *teleportation*. He conceived of this as a "transportory force" that could be used to explain everything from unexplained falls of rocks from the sky to poltergeist activity.

"In attempting to rationalize various details that we have come upon, or to assimilate them, or to digest them, the toughest meal is swallowing statements upon mysterious appearances in closed rooms, or passages of objects

2. Fort, *Complete Books of Charles Fort*, 480.

and substances through walls of houses, without disturbing the material of the walls," he wrote in *Lo!*.[3]

He goes on to explain, "The look to me is that, throughout what is loosely called Nature, teleportation exists, as a means of distribution of things and materials, and that sometimes human beings have command, mostly unconsciously, though perhaps sometimes as a development from research and experiment, of this force."[4]

Fort was also an early proponent of explaining many paranormal phenomena through the intervention of extraterrestrial beings, rather than using religion or spirituality.

In *New Lands,* published in 1923, Fort wrote, "Some day I shall publish data that lead me to suspect that many appearances upon this earth that were once upon a time interpreted by theologians and demonologists, but are now supposed to be the subject-matter of psychic research, were beings and objects that visited this earth, not from a spiritual existence, but from outer space. That extra-geographic conditions may be spiritual, or of highly attenuated matter, is not my present notion, though that, too, may be some day accepted."[5]

He speculated on the influence of otherworldly intelligences in the demonstration of anomalous phenomena,

---

3. Fort, *Complete Books of Charles Fort*, 488.

4. Fort, *Complete Books of Charles Fort*, 491–2.

5. Fort, *Complete Books of Charles Fort*, 367.

often in a tongue-in-cheek manner, to explain their baffling oddity.

Some would come to refer to this collectively as the "Cosmic Joker," although Fort himself referred to a sort of "godness" that he said "so much resembles idiocy that to attribute intelligence to it may be even blasphemous."[6]

It is to this inscrutable entity that he attributes the often nonsensical nature of the unexplained. This is something we see reflected today in the Trickster narrative so often applied to high strangeness concepts. That some strange being we do not—or perhaps cannot—understand is pulling the strings behind everything from UFOs to cryptid sightings to synchronicity in service to its own impenetrable motivations is becoming increasingly popular among researchers and investigators and was certainly an idea favored by Keel.

As Fort wrote in *Lo!*, "But if a godness places kindly lights in the sky, also is it spreading upon the minds of this earth a darkness of scientists. This is about the beneficence of issuing warnings and also seeing to it that the warnings shall not be heeded. This may not be idiocy. It may be 'divine plan' that surplus populations shall be murdered. In less pious terms we may call this maintenance of equilibrium."[7]

---

6. Fort, *Complete Books of Charles Fort*, 517.

7. Fort, *Complete Books of Charles Fort*, 673.

And finally, there is the underlying connectivity that Fort's works bring to the weird. The idea that, rather than placing ghosts and cryptids and UFOs and time slips and other such seemingly disparate phenomena in their own separate boxes, we should instead recognize their myriad commonalities.

"If there is an underlying oneness of all things, it does not matter where we begin, whether with stars, or laws of supply and demand, or frogs, or Napoleon Bonaparte. One measures a circle, beginning anywhere," he wrote in *Lo!*.[8]

And then, in *Wild Talents*, published in 1932, he wrote, "My liveliest interest is not so much in things, as in relations of things. I have spent much time thinking about the alleged pseudo-relations that are called coincidences. What if some of them should not be coincidences? [. . .] Sometimes I am a collector of data, and only a collector, and am likely to be gross and miserly, piling up notes, pleased with merely numerically adding to my stores. Other times I have joys, when unexpectedly coming upon an outrageous story that may not be altogether a lie, or upon a macabre little thing that may make some reviewer of my more or less good works mad. But always there is present a feeling of unexplained relations of events that I note, and it is this far-

8. Fort, *Complete Books of Charles Fort*, 468–9.

away, haunting, or often taunting, awareness, or suspicion, that keeps me piling on."[9]

Fort advocated for the shedding of dogmatic thinking and assumptive rationalizations, writing in *Wild Talents* that he could "conceive of nothing, in religion, science, or philosophy, that is more than the proper thing to wear, for a while."[10]

It is necessary to investigate unfettered by such philosophical weight, he argued, since these restrictive modes of analysis were preventing meaningful insight.

"Every science is a mutilated octopus," he said. "If its tentacles were not clipped to stumps, it would feel its way into disturbing contacts."[11]

And those contacts, naturally, are exactly what we are looking for.

Fortean phenomena aren't simply historical trivia to be mined from the archives and paraded around like some lost archaeological treasure. They are our constant companions—dwelling forever beside us, just out of reach but always close enough to intrude without warning. So it was that Forteana forced its way into the life of one Illinois woman in the spring of 2014.

---

9. Fort, *Complete Books of Charles Fort*, 733.

10. Fort, *Complete Books of Charles Fort*, 844.

11. Fort, *Complete Books of Charles Fort*, 866.

Paula contacted The Singular Fortean Society because she wanted to share a series of events that defied easy explanation.

The Singular Fortean Society was founded by Tobias and Emily Wayland as an investigative organization specializing in anomalous phenomena. Often, as was the case with Paula, witnesses would contact them first via email, which then leads to a phone interview. Tobias stayed in contact with Paula for years after their conversation, inviting her to participate in television and documentary productions related to the Mothman phenomenon.

On April 20, 2014, Paula was sitting near a window in the bedroom of her ground-level apartment near the Lakewood Forest Preserve in Wauconda when she saw a bright beam of light "come shooting down" across the street at around 11:00 p.m.

"It was so bright and defined," she said.

The light, which lasted between fifteen and twenty seconds, was an orange-gold color and did not illuminate its surroundings, nor did it move much, making only a slight "side-to-side or back-and-forth" motion.

The light retracted at one point, only to reappear a moment later.

Paula fell back onto her bed in shock.

"Oh my god, this thing sees me!" she remembered thinking. "I had the deepest feeling this thing saw me, it knows."

But the light retracted a final time and was gone.

"I heard a hum at the end, just before it took off," Paula said.

Three years later, in 2017, also in April, Paula encountered a terrifying winged being in the same area. She was taking the garbage out just before 6:00 a.m. when something drew her eye from across the street. In the early morning gloom, she saw a huge winged being.

Paula described the being as perhaps seven or eight feet tall when standing erect, but it carried itself hunched over, limping along. It was completely black, and from what she could see, its upper body and head were covered in hair. The being had long leathery wings, which were partially wrapped around its body as it moved toward Paula, jumping closer like someone moving through a strobe light. The terrifying creature made an unnatural groaning noise as it advanced.

Paula felt a palpable sense of evil emanating from the bat-winged monster.

"This is evil," she recalled. "I'm seeing evil."

Fearing for her safety, she quickly turned to flee into her apartment. After fumbling with her keys for a moment, she opened the door and turned around, terrified that the creature might be right behind her—but it was gone.

Paula said that she only knew a few of her neighbors personally, and unfortunately the ones she knew had not witnessed either event; although one of her neighbors did take her encounters seriously.

"I don't know what to say," Paula said of her experiences. "But I know what I saw."

These events, if viewed through the lens of strict categorization that has overtaken paranormal philosophy, might appear unrelated. But perhaps a different perspective is called for. Suppose for a minute that these phenomena might be, in some way, interdimensional, and then imagine what that could mean.

Think of a four-dimensional entity, for instance; to them, humanity would be as a two-dimensional creature is to us. This 4D being would be privy to an entire extra dimension through which it could interact with the universe. From a two-dimensional perspective, a three-dimensional being would be able to interact with their environment in ways that seem impossible.

Our 3D ability to, say, look down into a flat square and see (or even remove) its contents without penetrating its sides would be truly confounding to a 2D creature. Lacking access to the third dimension that makes this possible, they might struggle to even conceive of any mechanism through which it could be done. And so might we struggle to grasp the machinations of higher-dimensional beings.

Consider how this might apply to a four-dimensional being. Since time is widely agreed upon to be the fourth dimension, then what does it mean for an entity to view it as we do three-dimensional space? What becomes possible when they can travel across time like we would the three dimensions surrounding us? In the case of Paula, it could mean that two events separated by years of time from our perspective occurred much closer together according to the mysterious beings responsible for them.

For all we know, from their point of view, the anomalous phenomena in question could have seemed no farther separated than two things happening in different corners of the same room would seem to us.

If this seems weird and confusing, fear not, for you are in good company.

THE MOTHMAN PROPHET

John Alva Keel, born Alva John Kiehle on March 25, 1930, was an adventurer and journalist perhaps best known in paranormal circles for introducing the world to Mothman. In the 1950s, during the Korean War, he was drafted into the United States Army, where his background in radio and television earned him an assignment to the American Forces Network in Frankfurt, Germany. Keel would later write in *Operation Trojan Horse* that he had been "trained in psychological warfare during [his] stint as a propaganda writer for the U.S. Army."[12]

Following his service in the military, Keel spent time wandering the Middle East and Asia, where he reportedly sought out everything from fakirs and yogis to the elusive Yeti. These travels would later be recorded in his book *Jadoo*, published in 1957. From there, he came back to the States where he worked in television, wrote novels under the pseudonym Harry Gibbs, and even almost had an article on UFOs published in *Playboy* (although they ultimately opted to publish one by Dr. J. Allen Hynek instead). And finally, in 1967, while investigating UFO sightings in West Virginia, Keel became embroiled in the Mothman phenomenon.

But before we turn our attention to that instance of high strangeness, it's important to first understand the philosophy present in Keel's work.

Keel wrote in his 1975 book *The Mothman Prophecies*, "Back in the 1920s, Charles Fort, the first writer to explore inexplicable events, observed you can measure a circle by beginning anywhere. Paranormal phenomena are so widespread, so diversified, and so sporadic yet so persistent that separating and studying any single element is not only a waste of time but also will automatically lead to the development of belief. Once you have established a belief, the phenomenon adjusts its manifestations to support that

---

12. Keel, *Operation Trojan Horse*, 267.

belief and thereby escalate it. If you believe in the devil, he will surely come striding down your road one rainy night and ask to use your phone. If you believe that flying saucers are astronauts from another planet they will begin landing and collecting rocks from your garden."[13]

Later, in that same work, he wrote, "Belief is the enemy."[14]

We see this reflected in the material quoted by Keel, but also in another quote from Fort in which he addresses the ephemeral nature of belief.

"Fashions often revert, but to be popular they modify. It could be that a re-dressed doctrine of witchcraft will be the proper acceptance. Come unto me, and maybe I'll make you stylish. It is quite possible to touch up beliefs that are now considered dowdy, and restore them to fashionableness. I conceive of nothing, in religion, science, or philosophy, that is more than the proper thing to wear, for a while," Fort wrote in *Wild Talents*.[15]

In fact, Fort's influence is something Keel wore proudly, even beginning each part of his book *The Eighth Tower*—a follow-up to *The Mothman Prophecies*, which included much of what had been left out of that work—with quotes from him.

---

13. Keel, *The Mothman Prophecies*, 8.

14. Keel, *The Mothman Prophecies*, 30.

15. Fort, *Complete Books of Charles Fort*, 844.

It is within *The Eighth Tower* that Keel explores the idea of a hidden world of raw energy he refers to as the Superspectrum.

"And there are, as we shall see, forms of energy on such high frequencies they cannot be detected with even the most sophisticated scientific instruments," Keel wrote. "If your eyes were tuned beyond the very narrow confines of the spectrum of visible light, you would find yourself looking into a thick fog of dazzling, unreal colors. Some psychics and UFO percipients have described these occult colors, and they have always been used to symbolize the supernatural entities. If you could peer into this Superspectrum, you would undoubtedly see some frightening things—strange shapes and eerie ghostlike forms moving through a sea of electrical energy like fish in some alien sea."[16]

This Superspectrum is home to another of Keel's most famous contributions: the Ultraterrestrials.

Of these beings often associated with UFOs, Keel wrote in *Operation Trojan Horse*, "There is a superabundance of historical documentation which plainly indicates that these objects and their elusive occupants have always been a part of the environment of earth and that they seem to know everything about us, are able to speak our languages, and

---

16. Keel, *The Eighth Tower*, 21.

are even familiar with the total lives of some—if not all—human beings."[17]

It is through the Superspectrum that the Ultraterrestrials are capable of manifesting in such a dazzling array of appearances, Keel opined. The qualities of that energy field also dictate how these entities seem to reflect our own subconscious beliefs back at us.

"When the levels of energy in this field are changed or somehow influenced by us, the whole character of these Superspectrum entities is altered," he wrote in *The Eighth Tower*. These energies, Keel noted, "often work against man because they affect our minds rather than our machines."[18]

Which, of course, explains why our technology is seemingly incapable of detecting any of this.

The Superspectrum and its resident Ultraterrestrials are used by Keel to explain everything from UFO occupants to ghostly phenomena.

"Just as the UFO cultists have settled upon the extraterrestrial thesis to explain flying saucers, the occultists have decided that poltergeists are caused by energy radiated from disturbed children or by restless ghosts," he wrote in *Operation Trojan Horse*. "The common procedure for investigating poltergeists is to examine the entire history of the house and land. In some cases, it is discovered that someone

17. Keel, *Operation Trojan Horse*, 60–1.

18. Keel, *The Eighth Tower*, 21.

was murdered there or buried there years—even hundreds of years—before the manifestations began. So the ghost is blamed for the phenomenon."[19]

However, he went on to say, "There are, of course, thousands of murders and violent deaths every year, and it might be useful to perform a study of all the scenes of these crimes to determine the percentage of hauntings that occur afterward. I suspect that percentage would be quite small."[20]

Instead, proposed Keel, "It seems that these events are concentrated in base areas and are merely a side effect of the other things taking place there unnoticed. Small, confined areas all over the world appear to be haunted century after century by mischievous entities who are able to adopt any guise and who maintain such total control over material objects that they can produce any kind of manifestation. Our willingness to accept the restless ghost theories is based upon our ego-inspired need to believe in the immortality of the human soul or spirit. The ultraterrestrials might recognize that need and wickedly take advantage of our beliefs, tailoring their manifestations so that they appear to support our religious convictions."[21]

If true, this would be a conspiracy of truly cosmic proportions, and what seems to hold it all together is synchronicity.

---

19. Keel, *Operation Trojan Horse*, 213.

20. Keel, *Operation Trojan Horse*, 213.

21. Keel, *Operation Trojan Horse*, 213.

"This is the law of synchronicity," Keel explained in *The Eighth Tower*. "Duplicate paranormal events occur simultaneously in different locations to lend credibility to a belief or frame of reference. Ufological lore abounds with synchronous events, but so, too, does religious, occult, and psychic lore. Paranormal incidents take place in massive waves, sometimes grouped years apart. Poltergeist manifestations erupt simultaneously with UFO flaps. Angels, demons, hairy monsters, and sea serpents all surface at the same time that UFOs are stopping lone drivers on remote back roads."[22]

These pockets of synchronicity, Keel said, often appear in "window areas." Window areas are geographical locations where "unusual events have occurred over and over again, century after century."

A sudden influx of anomalous activity within a window area is known as a flap. Flaps of activity occurring within these window areas, he said, are explained erroneously by ufologists who point out "that it takes a long time for the spaceships to go back and forth from their home planet," and by occultists as the earth passing through "zones in space inhabited by terrible spiritual beings. Each time we enter such a zone, the beings ooze through holes in our 'etheric envelope,' these holes being located in what I call window areas."

---

22. Keel, *The Eighth Tower*, 117.

All of which seems strongly reminiscent of the godness that Fort said, "So much resembles idiocy that to attribute intelligence to it may be even blasphemous."

Author and researcher Brent Raynes, who corresponded with Keel for decades prior to Keel's death in 2009, credits the late investigator with shifting his perspective away from the nuts-and-bolts hypothesis for UFOs.

He also graciously agreed to be interviewed for this book.

After reading Keel, Raynes said, "I realized that the phenomenon was potentially far more complex and involved far more components and processes than I had initially thought."

To Raynes, Keel was ahead of his time.

"There were a number of areas that he really shined in, but much of the mainstream wasn't very welcoming to his ideas," he explained. "They were pretty (and still are) entrenched within only what applies to the perceived ET nuts-and-bolts theory. Keel's fieldwork, I do believe, helped guide him in understanding the more complex aspects of all of this, and later, when I discovered he'd had anomalous experiences in his childhood and younger years (an interview done of Keel by Tim Beckley brought that to my attention), it made sense to me that uncovering what he did in the field would also come to click with him based on seeing a UFO and dealing with a poltergeist at a young age."

Keel often struggled to make sense of high strangeness, especially in the early days of his career, although he did seem to believe there was something to it.

"In the beginning I feel Keel struggled and had to learn how to try and best navigate the complex elements of close-range UFO sightings, landings, and occupant encounters, the 'missing time' cases, the contactees/abductees, the MIB, cryptids, and all the psychic aspects that often turned up in these cases," Raynes said.

"Early on, he got spooked and fell victim to that 'emotional quicksand' that he warned [about]. He warned fellow researchers about alien gas attacks (he wrote an article about UFO mystery gas attacks in Saga), I have been told Keel had purchased gas masks himself and warned fellow researchers to get their own gas masks. His actions generated some great fear and concern among fellow researchers.

"There was a time he was looking for actual alien bases. Over time he realized though that a lot of this activity isn't what it appears to be. The activity can lead you on wild-goose chases. You can end up chasing unreal chimeras. It can be a real theater of the mind."

And that realization brought about some of Keel's most groundbreaking work, according to Raynes.

"Recognizing the potential for other areas to be interrelated [is one of Keel's greatest contributions to the field of anomalous studies]," Raynes said. "He felt ufology should be a branch of parapsychology. He pointed out how ghosts and aliens, is there really a significant difference, beyond the frame of reference? Both can float, walk through solid walls, communicate telepathically. Do many of the same things.

"Tim Beckley congratulated him ... with waking a global ufological audience to the psychic elements of the UFO phenomenon, and Keel replied that it was like opening Pandora's box, really. It only made the problem all the more complicated.

"A few years before his passing, Keel wrote to me of how he was unsuccessful in making a solid case for his Ultraterrestrial theory, but he strongly felt that it was the most probable explanation for all of the high strangeness elements of the UFO phenomenon."

Now, having more or less caught up with the gist of Keel's perspective, we might consider ourselves sufficiently girded in weirdness to enter Mothman country.

We'll begin our tour of the Ohio River Valley shortly, but before we do, let's make a short pit stop in Parsonsburg, Maryland, where a woman named Bobbi said she and her son encountered a mysterious winged humanoid in May of 2019.

Later that year, she would submit a report to The Singular Fortean Society.

"We had gone to Walmart on a Friday night. We're night owls, so we always do late-night 'oh, let's run here, let's run there,'" Bobbi said. "It was probably between midnight and 1:00 a.m. We were on Dagsboro Road, and it cuts from town to the country, basically. We live in Parsonsburg, and Walmart is in Salisbury. It's a very winding road, like very, very twisty."

She was driving with her son in the passenger seat, cruising at only thirty miles per hour to better navigate the winding country road. As she rounded a curve, Bobbi said she saw something twenty feet or so ahead in the road.

"It was crouched, and I had a couple of thoughts. The first one was, 'It's going to hit my car,' then the second thought was, 'What the hell kind of bird is that out here at 1:00 a.m. in the morning?' Then, when I got closer, I saw that it didn't look like a bird. Its skin was smooth, and my brain wasn't registering it at all. By the time I was on it, it had started to stand up and was probably six feet tall, at least."

Bobbi's mind reeled at the seemingly impossible situation unfolding before her.

"It was big, you know? I was sure it was going to hit the front of my car or something. So, I slowed down a little more and kept going, and as I got closer to it, it started to move, and I guess I was just staring at it, and it seemed to be crouched down—like a buzzard or something. When it raised up, I didn't see anything on the road, I didn't see anything that it was crouched eating, like a buzzard would. It lifted itself up, and I honestly to this day cannot even tell you what I saw, until it reached the tip of my windshield.

I still was thinking, 'This thing is going to hit my car!' Then, as it went up, and fully stretched its body out, I could see that it was very gray and very smooth in appearance, like it would be smooth if I touched it. It was bald, it was bald-headed, and I saw feet, but like human feet. Like gray human feet.

"It was crouched on the left side of the road. It went up straight and then over across to the right where my son was and then up over the trees on that side. He saw the front from a side angle as it flew over us. The last thing we saw was the feet. And then he was just gone. We didn't see him glide over the trees; he was just gone.

"It seemed to be very, very bald. The top of the head was bald. It was very round and very bald. There were ears. I thought it was a human head," she explained.

According to her son, who had seen more of the creature's face, it "looked like a skull, with sunken eyes, and with skin over it."

It took Bobbi a moment to react before she looked to her son for confirmation of what she'd seen.

"I just stared; I really didn't process what I was seeing. It flew straight up in front of my car, and when its feet were up above my car is when I was like, 'Holy shit!' It went up over the trees, it never flapped its wings—it just kind of glided up off the ground. I waited for a second, and then I looked at my son and I asked, 'Did you see that? Am I crazy?' His face was completely still. He had no expression, and all he said to me was, 'I saw bat wings and human feet.' And then, we didn't say another word about it," she said.

Both Bobbi and her son seemingly forgot about the event completely until several weeks later when they were driving down the same stretch of road at night.

"We drove home in a stunned daze, and we didn't say anything else," she said. "And then a couple of weeks later we were driving down that road again and I said [to my son], 'Oh my god, do you remember what we saw?' And he said, 'Yes, oh my god!'"

The two searched online but couldn't find anything similar to what they'd seen—until Bobbi's son came across a video about Mothman on YouTube.

"He found some stuff on the Mothman. We started googling Mothman stories, and that's when we came across [The Singular Fortean Society]," she said.

Bobbie was struck by the number of sighting reports similar to their own.

"That's when I was like, 'Holy crap!' Because when I saw the sketches [by artist Donie Odulio] that were on one of the stories, it matched to a T [what we had seen]. Besides the eyes, we didn't see any red eyes, but I didn't really see any specific features. My son said he saw the face, and he said, 'No, the eyes were just kind of sunk in,' and he said he saw wings and hands."

Although her son said he has a strong feeling of fear associated with the encounter, Bobbi said she didn't experience any overpowering emotions, other than bewilderment at how they so suddenly and mysteriously forgot the event.

"We're very talkative, and we love creepy stuff, but we'd never seen anything," she said. "That's what threw me for a loop, because we literally just drove home after in silence. I didn't wake up the next day and tell a single soul. I didn't even remember it until we drove back down that road."

Following the event, the family experienced an incident of a potentially parapsychological phenomenon that they didn't want shared publicly. In addition to that, according to Bobbi, ever since the event, their "kitchen radio turns on by itself all the time." The radio was purchased before the incident and had no prior history of malfunction.

Both Bobbi and her son otherwise handled their experience well, and more than anything were relieved to find that they weren't alone.

"Other people have seen this," Bobbi said. "We're not crazy."

Bobbi and her son are as sane as anybody, which we can just as easily say about everyone we're about to meet. Now, on to Point Pleasant.

CHAPTER 2
BIRTH OF A LEGEND

The world was officially introduced to Mothman by an article published in the *Point Pleasant Register* on Wednesday, November 16, 1966, entitled "Couples See Man-Sized Bird ... Creature ... Something."

The article told of two young couples from Point Pleasant, West Virginia—Roger and Linda Scarberry and Steve and Mary Mallette—who said that they had seen a large black creature with glowing red eyes in their car headlights the night before while visiting a nearby abandoned TNT depot. This depot, commonly known as the TNT area, has since become so firmly ingrained in the Mothman legend that it remains a popular spot for monster hunting to this day ... but more about that later.

The couples' ordeal began between 11:30 p.m. and midnight when they spotted something unusual near an old power plant adjacent to several National Guard Armory buildings. According to a description given to the press, what they witnessed was a man-size, birdlike creature that stood about six or seven feet tall, with a ten-foot wingspan and glowing red eyes. The eyes were said to be approximately two inches in diameter and six inches apart.

"It was like a man with wings," Steve said. "It wasn't like anything you'd see on TV or in a monster movie."

Although it looked "like a man with wings," its head was "not an outstanding characteristic."

Furthermore, the winged oddity "didn't resemble a bat in any way," but "maybe what you would visualize as an angel."[23]

Roger would later tell John Keel the creature "was shaped like a man, but bigger. Maybe six and a half or seven feet tall. And it had big wings folded against its back."

To that, Linda added, "But it was those eyes that got us. It had two big eyes like automobile reflectors."

"They were hypnotic," Roger said of the thing's eyes. "For a minute we could only stare at it. I couldn't take my eyes off it."

Understandably afraid, the quartet decided to beat a hasty retreat.

---

23. "Couples See Man-Sized Bird ... Creature ... Something."

"I'm a hard guy to scare," Roger said, "but last night I was for getting out of there."

Although it first seemed to slowly turn and shuffle toward the power plant's door, which hung off its hinges, the creature—or another one just like it—was again spotted by the astonished witnesses on a hill overlooking the exit road upon which they were making their escape. This time it gave chase as they fled. Reportedly a clumsy runner, the creature was fast in flight, managing to keep pace with their car at speeds up to one hundred miles per hour.

"We were doing one hundred miles an hour and that bird kept right up with us," Roger told Keel. "It wasn't even flapping its wings."[24]

As it flew overhead, Mary said, "It squeaked like a mouse."

"Funny thing, we noticed a dead dog by the side of the road there. A big dog. But when we came back a few minutes later, the dog was gone," Roger added.

The thing seemed to glide overhead until they reached the National Guard Armory on Route 62, at which point it fled into a nearby field.

Roger and Steve said the creature's eyes only glowed red when their lights shined on it, and it seemed to want to get away from the lights.

---

24. Keel, *The Mothman Prophecies, 39.*

"It apparently is afraid of light," Steve speculated, "and maybe it thought it was scaring us off."

After that, they headed back to Point Pleasant to regain their courage before again setting out in search of the monster.

"We went downtown, turned around, and went back, and there it was again," Steve said. "It seemed to be waiting on us."[25]

They saw the creature three times throughout the night, the final encounter occurring at the gate of the C. C. Lewis farm on Route 62. There was a sound of wings flapping and the creature took off straight up, like a helicopter.

Both couples were so terrified by what they had seen that they went to the police. Sheriff's deputies and city police searched the area at about 2:00 a.m. that morning but came up empty-handed.

Although authorities found no definitive evidence of the winged humanoid, Deputy Millard Halstead said he had seen dust being kicked up in the vicinity of a coalfield that "could have been" caused by the creature.

He also reportedly experienced a "very loud signal" blasting from his police radio when he tried to turn it on in the TNT area.[26]

---

25. Keel, *The Mothman Prophecies, 39.*

26. Hyre, "Winged, Red-Eyed 'Thing' Chases Point Couples Across Countryside."

The sound, Keel wrote, "was a loud garble, like a record or tape recording being played at very high speed."

Furthermore, Halstead would tell Keel, "I've known these kids all their lives. They'd never been in any trouble and they were really scared that night. I took them seriously."[27]

None of the witnesses had an explanation for the bizarre being but speculated that it might live in the abandoned power plant, possibly in one of its large boilers.

"There are pigeons in all the other buildings," Steve said, "but not in that one."

Ultimately, though, this experience was without any definitive answers.

"If I had seen it while by myself, I wouldn't have said anything," Roger said, "but there were four of us who saw it."

"This doesn't have an explanation to it. It was an animal but nothing like I've seen before," Steve added.[28]

The next morning, Sheriff George Johnson called a press conference, during which the Scarberrys and Mallettes were interviewed by members of the media. Mary Hyre, a reporter for the *Athens Messenger* and a key player in events surrounding the anomalous phenomena experienced in and around Point Pleasant, would send the story out on the *Associated Press* wire that day, alerting the surrounding area to the monster's purported presence. According to Keel,

---

27. Keel, *The Mothman Prophecies*, 39.

28. "Couples See Man-Sized Bird ... Creature ... Something."

this led to "some anonymous copy editor" giving the monster a name "spun off from the Batman comic character who was then the subject of a popular TV series."[29] That name, of course, was Mothman.

This sighting and its resulting press conference excited the public's imagination about the possibility of a new monster haunting the Ohio River Valley, but as time would tell, it may not have been humanity's first encounter with such a creature in West Virginia.

That honor might instead belong to an anonymous national guardsman.

As Keel wrote in *The Mothman Prophecies*, on November 1, 1966, "A national guardsman was working outside the national guard armory on the edge of Point Pleasant when he saw a figure perched on the limb of a tree beyond the high fence. At first he thought it looked like a man, but after he studied it for awhile he decided it was some kind of bird. The biggest bird he had ever seen. He went to call some friends and when they came the bird was gone."[30]

Another better-documented sighting also took place before the Scarberrys' and Mallettes' fateful encounter with a "man-sized bird… creature… something" in Point Pleasant's TNT area.

---

29. Keel, *The Mothman Prophecies*, 39.

30. Keel, *The Mothman Prophecies*, 35.

In an article for the *Charleston Gazette* published Friday, November 18, 1966, Kenneth Duncan of Blue Creek in Kanawha County described his sighting of something that "looked like a brown human being" in a wooded area at Reamer near Clendenin.

Kenneth said that he and four other men were digging the grave of his father-in-law, Homer Smith, also of Blue Creek, at the time of his sighting.

"It was gliding through the trees and was in sight for about a minute," Duncan said of the creature.

Unfortunately for Kenneth, the other men he was with—Robert "Bob" Lovejoy of Allen, Michigan; William "Bill" Poole, also of Allen, Michigan; Andrew Godby of Blue Creek; and Emil Gibson of Quincy—didn't look up in time to see the creature, whatever it was.[31]

While these experiences are certainly disturbing enough, perhaps none of them can be said to be as terrifying as that described by Marcella Bennett.

Following Sheriff Johnson's press conference, curious folks from Point Pleasant flocked to the TNT area. Cars lined the roads between abandoned buildings and men with guns searched high and low for any sign of the monster. This exaggerated interest, Keel thought, could be due to Point Pleasant being "a town of six thousand people,

---

31. "Eight People Say They Saw 'Creature.'"

twenty-two churches, and no barrooms"; something that made Mothman "almost a welcome addition."[32]

During this town-wide monster hunt, Keel said, "A large red light moved around in the sky directly above the TNT area." This light went unnoticed by most, but not by Marcella Bennett.

Marcella was in a car with her infant daughter, Tina, and her friends Mr. and Mrs. Raymond Wamsley when they saw the light fly overhead.

"It wasn't an airplane. We couldn't figure out what it was," Marcella told Keel.

The group wasn't particularly interested in monster hunting. They were on their way to visit Ralph and Virginia Thomas, who just happened to live in a bungalow set back among the TNT area's empty igloos.

Virginia, Keel noted, possessed "second sight."

"She had accurately predicted numerous accidents and local events over the years," he wrote. "She was careful not to seek attention and only her friends knew of her remarkable abilities."[33]

Being a religious sort, Virginia went to church most evenings, and so it was that she and Ralph were out when Marcella, Tina, and the Wamsleys dropped in.

---

32. Keel, *The Mothman Prophecies*, 40.

33. Keel, *The Mothman Prophecies*, 40.

Three of the Thomas children were home, and they spoke with them briefly before turning to leave. It was then that they noticed a dark figure emerging from the shadows behind their parked car.

"It seemed as if it had been lying down," Marcella said of the figure. "It rose up slowly from the ground. A big gray thing. Bigger than a man, with terrible glowing red eyes."

As Keel wrote, Marcella "uttered a little cry, so horrified she dropped the small baby in her arms. The child began to cry, more insulted than hurt, but her mother couldn't move to pick her up again. She stood transfixed, hypnotized by the blazing red circles on the top of the towering, headless creature. Its great wings unfolded slowly behind its back."

Luckily, Raymond Wamsley was not so transfixed and grabbed the paralyzed Marcella and fallen Tina. They then ran back to the house, slamming and locking the door.

Just then, there was a sound on the porch, and they saw two red eyes peeking into a window. At that point, according to Keel, the "women and children became hysterical while [Raymond] phoned the police."[34]

But by the time the police arrived, the monster had gone.

Unfortunately for Marcella and the other residents of Point Pleasant, this was just the beginning.

---

34. Keel, *The Mothman Prophecies*, 40.

CHAPTER 3

KEEL GOES FULL FORTEAN

Keel, of course, was a Fortean and liked making connections between anomalous phenomena, and when you have a phenomenon as anomalous as Mothman, the natural proclivity for someone like that is to connect it to everything. Take the tale of Newell Partridge, for example.

Newell's dog, a German shepherd named Bandit, had gone missing on November 14, 1966, just over a day before the Scarberrys and Mallettes had their encounter in the TNT area.

"It was about 10:30 that night, and suddenly the TV blanked out," Partridge told investigator Gray Barker, according to Keel. "A real fine herringbone pattern appeared on the tube, and at the same time the set started a loud whining noise, winding up to a high pitch, peaking and breaking off, as if you were on a musical scale and you went as high as you could and came back down and repeated it. […] It sounded like a generator winding up. It reminded me of a hand field generator that one might use for portable radio transmission in an emergency."

It was then that Newell heard Bandit howling out on the porch. Grabbing a flashlight, he went to investigate.

"The dog was sitting on the end of the porch, howling down toward the hay barn," Partridge continued. "I shined the light in that direction, and it picked up two red circles, or eyes, which looked like bicycle reflections. Still there was something about those eyes that is difficult to explain. When I was a kid I night-hunted all the time, and I certainly know what animal eyes look like—such as coon, dog, and cat eyes in the dark. These were much larger for one thing. It's a good length of a football field to that hay barn. Probably about 150 yards; still those eyes showed up huge, for that distance."[35]

When Newell's flashlight found those red "eyes," Bandit took off after them. Overcome with fear, Newell stayed frozen in place and watched as his dog disappeared into the darkness.

That night, Newell reportedly slept with a gun next to his bed, and the next morning, his courage recovered in the daylight, he set out to find Bandit.

---

35. Keel, *The Mothman Prophecies*, 36.

"I walked out to the barn, looking for tracks. Here and there I could see Bandit's paw prints. These were rather easy to find, for he was a heavy dog, and the area was muddy," he said.

Once Bandit's tracks reached the approximate position of the anomalous red "eyes," Newell noticed that they seemed to go in a circle, as though the dog had been chasing his tail. Although, Newell noted, "he never did that."

"And that was that," Newell said of the experience. "I couldn't see them go off anywhere, though I did see a series of fresh tracks which apparently led from the porch to the spot where he ran in circles. There were no other tracks of any kind."

Bandit had seemingly vanished.

For Newell, part of the experience he struggled with the most—as is often the case with witnesses—was how it made him feel.

"I think that the hardest thing to explain is the feeling involved [...] except to say it was an eerie feeling," he said. "I have never had this sort of feeling before. It was as if you knew something was wrong but couldn't place just what it was."[36]

Naturally, Bandit's disappearance was compared to the dead dog later mentioned by Roger Scarberry during that group's encounter with the winged being in Point Pleasant,

---

36. Keel, *The Mothman Prophecies*, 37.

and as the narrative spun by Keel in *The Mothman Prophecies* became more popular, many would begin to ask themselves if these otherworldly beings might be dangerous.

Strange incidents that could be tied to the winged humanoid sightings at that time were plentiful and Keel was eager to include them in his Ultraterrestrial narrative, which as mentioned earlier is expansive enough to fit almost anything. Because of this, Mothman certainly seemed to have plenty of weird company in the Ohio River Valley during the mid to late 1960s, and not the least among them were UFOs and their eerie, elusive occupants.

It wasn't unusual for Mothman witnesses to have one or more experiences with UFOs, too, like the one described above the TNT area by Marcella Bennett. Keel himself claimed to have seen UFOs during his time in Point Pleasant. One night in November of 1967, after interviewing local UFO witnesses with Mary Hyre, he wrote in *The Mothman Prophecies*, "We ourselves had watched a very strange light in the sky. Since there was a heavy, low cloud layer, it could not have been a star. It maneuvered over the hills, its brilliant glow very familiar to both of us for we both had seen many such lights in the Ohio valley that year."[37]

UFO sightings seemed to be a constant companion to Mothman in and around Point Pleasant. After one incident in which Keel and Mary Hyre reportedly spotted a "large

---

37. Keel, *The Mothman Prophecies*, 5.

black object" following a plane, they returned to Hyre's office to find all the phones ringing.

"People were seeing flying objects all up and down the valley," he wrote.

Keel noted that this UFO sighting flap took place on a Wednesday—something to which he gave great significance—attributing it to something he called the Wednesday phenomenon.

Namely, that out of some seven hundred UFO reports he had collected and analyzed, the greatest number of sightings, 20 percent, had taken place on Wednesdays.

"The Wednesday phenomenon works. I've been studying it for years and I still can't say why it works," Keel wrote. "Researchers in other parts of the world have now followed my example and found similar time patterns in the sightings of their own countries."

He even provided a table in *The Mothman Prophecies* that featured "a breakdown of sightings recorded in 1950," showing reported UFO sightings for that year from the United States, Spain, and Belgium. According to the data provided by Keel, Wednesdays hosted the most sightings with 18.8 percent of sightings reported on that day, followed by Thursdays with 16.5 percent, and Mondays and Tuesdays with 14.7 and 14.6 percent, respectively.[38]

---

38. Keel, *The Mothman Prophecies*, 89.

These sightings weren't limited to just unidentified craft; sometimes, witnesses claimed to interact with their occupants too. Perhaps one of the strangest encounters, and certainly the best known among those associated with Mothman sightings during this period, was that of Woodrow Derenberger and the spaceman known only as Indrid Cold.

At 7:00 p.m. on November 2, 1966, Woodrow was headed home in his truck on Interstate 77 outside of Parkersburg, West Virginia, when he noticed what he at first thought to be a car overtaking him on the road. He had just been passed by another car and decided to slow down a bit; he was speeding slightly and thought that perhaps this second vehicle might be a police officer. But when this vehicle pulled past and swerved in front him, slowing down and turning to block the road, Woodrow quickly realized this was no terrestrial form of transportation.

As he slammed on the brakes to avoid a collision, stopping less than ten feet short of a collision, he saw that the craft was charcoal gray and shaped like "an old-fashioned kerosene lamp chimney, flaring at both ends, narrowing down to a small neck and then enlarging in a great bulge in the center."

Woodrow stared in disbelief as a door opened in the side of the craft and a man stepped out into the road.

The man stood about five feet ten inches tall and had long dark hair combed straight back. The heavily tanned skin of his face was split by a huge grin as he stood with his arms crossed and his hands tucked under his armpits. Underneath a dark topcoat, Woodrow saw what appeared to be some kind of garment made of "glistening greenish material" that was "almost metallic in appearance."

Silently, the stranger approached the truck.

"I didn't hear an audible voice," Woodrow later told Keel. "I just had a feeling ... like I knew what this man was thinking. He wanted me to roll down my window."

Woodrow sensed the mystery man communicate soundlessly, "Do not be afraid. We mean you no harm. I come from a country much less powerful than yours."

He asked for Woodrow's name, and when given it, introduced himself only as Cold.

"I sleep, breathe, and bleed even as you do," Cold explained to Woodrow.[39]

Nodding toward Parkersburg, Cold asked Woodrow what kind of place it was.

Woodrow explained that it was a city, to which Cold replied that, in his world, such places were called gatherings.

Cars continued to drive by the duo during this telepathic conversation, seemingly not noticing the lamp-shaped craft as it hovered forty or fifty feet above the road.

---

39. Keel, *The Mothman Prophecies*, 33.

After several minutes of baffling conversation, Cold told Woodrow to report his experience to the authorities, promising that he would come forward later to corroborate it. Announcing then that he would see Woodrow again soon, Cold reentered the craft as it descended to the road, and just as quickly as it had come, the mysterious ship and its otherworldly occupant departed silently into the night sky.

Woodrow was unusually lucky in two respects, outside of the infinitesimally small chances of having this kind of encounter in the first place, in that he had a supportive family, and further witnesses came forward to corroborate his account.

Seeing how upset he was upon arriving home, Woodrow's wife insisted that he call the Parkersburg police to report his encounter.

After being thoroughly questioned by city and state police, Woodrow's story began to circulate in the local press. Appearing on radio and television, Woodrow became a local celebrity. Throngs of people arrived at Woodrow's home, hoping to witness a UFO, and his phone began to ring off the hook.

Also during this time, according to Keel, "People who had driven that same route the night before came forward to confirm that they had seen a man speaking to the driver of a panel truck stopped on the highway. Mrs. Frank Huggins and her two children had reportedly stopped their own car and watched the object soar low over the highway minutes after Woody watched it depart. Another young man said the object had frightened him out of his wits when it hovered over his car and flashed a powerful, blinding light on him."[40]

Amidst the tumult, Woodrow would be again visited by Cold, who this time showed up at his home in a black Volkswagen. Woodrow and Cold spoke casually at the edge of his porch for a few minutes, and before he left, Cold left Woodrow with two things: a vial of medicine to cure Woodrow of a chronic stomach ailment, and his first name—Indrid.

Woodrow claimed that his life grew increasingly bizarre after these events. He said that he was driving with a coworker along Route 7 on November 4, just two days after his initial encounter, when he received a telepathic message from Cold. In it Cold explained that he was from the utopian planet Lanulos in the galaxy Ganymede, and that Lanulos was much like Earth. Cold said that he had a wife, Kimi, and two sons, adding that the people of Lanulos had life expectancies of 125 to 175 years.

All this otherworldly attention apparently attracted the curiosity of our own government. Woodrow said that he and his family were taken to Cape Kennedy where he was questioned at length about his interactions with Indrid Cold.

40. Keel, *The Mothman Prophecies*, 34.

His family, too, would reportedly meet Indrid and other beings who supposedly came from a planet called Lanulos. Woodrow's wife, Keel said, was "frightened of them and felt they were engaged in something evil." Furthermore, she would tell Keel, these beings were just like us, traveled in ordinary automobiles, and "were probably infiltrating the human race in large numbers."[41]

These events would come to include trips to the Ganymede galaxy with Indrid, Woodrow claimed. The trips would seemingly take many hours or days, but always upon his return, Woodrow would find that only a few hours had passed on Earth.

Amidst all the high strangeness, Woodrow would receive strange phone calls, despite having switched to an unlisted number. These calls included threats against him if he didn't shut up about his encounters with Cold, and some that consisted only of "eerie electronic sounds and codelike beeps."

Woodrow wasn't alone in the seemingly paranormal harassment he received. Several residents of West Virginia and Ohio reported receiving strange phone calls from someone speaking in a "metallic voice," among other anomalies. One such person happened to be the daughter-in-law of Mr. and Mrs. James Lilly.

---

41. Keel, *The Mothman Prophecies*, 70.

At the same time, the Lilly family was reportedly plagued by unsettling phenomena at their home on Camp Conley Road near Point Pleasant's TNT area.

"It didn't take us long to learn that when our TV started acting up it was a sure sign that one of those lights was passing over. I didn't think much of all the flying saucer talk until I started seeing them myself. You've got to believe your own eyes," said James Lilly, noted by Keel to be a "no-nonsense riverboat captain."

The Lillys kept their UFO sightings to themselves, but it wasn't long before word spread and soon their street was filled with seekers hoping to catch a glimpse of something otherworldly.

"We've seen all kinds of things," James's wife Jackie told Keel. "Blue lights, green ones, red ones, things that change color. Some of them have been so low that we thought we could see diamond-shaped windows in them. And none of them make any noise at all."

Soon other bizarre occurrences intruded into their lives.

Cars on the street near their house inexplicably stalled—a phenomenon commonly associated with UFO sightings and reported by other recipients of unusual activity in the area around the same time. Their kitchen cabinets would slam in the dead of night, and at least once their living room door, which was secured with both a chain and a snap lock, was found open in the morning. The family reported loud metallic sounds, "like a pan falling," and Jackie said she had heard a phantom baby crying.

"It sounded so plain that I looked around the house even though I knew there was no baby here," she said. "It seemed to come from the living room … only a few feet away from me."[42]

Furthermore, James and Jackie's daughter Linda had experienced a singularly disturbing visitation amidst the clamor of poltergeist activity.

When asked by Keel if she had ever dreamed of a stranger in their house, she reluctantly admitted that she had awoken one night to find a shadowy figure looming over her bed.

"It was a man," she said. "A big man. Very broad. I couldn't see his face very well but I could see that he was grinning at me."

Jackie added that she heard her daughter scream but thought that she was only having a bad dream.

"Jim was working on the river that night. And Linda woke me up with a terrible scream," Jackie said. "She cried out there was a man in her room. I told her she was dreaming. But she screamed again."

The man approached Linda and she noticed he was wearing a checkered shirt.

---

42. Keel, *The Mothman Prophecies*, 77.

"He walked around the bed and stood right over me," she said. "I screamed again and hid under the covers. When I looked up again, he was gone."

At that point, Linda fled from her bedroom to seek solace with her mother.

"She came running into my room," Jackie said. "She cried, 'There is a man in my room! There is!' She's refused to sleep alone ever since."[43]

Many of the details of these events would come to be echoed decades later and hundreds of miles away by a family in Madison, Wisconsin, who had lived through their own infestation of high strangeness.

The Singular Fortean Society was contacted in early 2020 by a woman who said she was looking for "assistance on finding answers to an incident my mom and I had happen about twenty years ago."

This woman, Selina, had been referred to the paranormal organization by Adam Benedict of The Pine Barrens Institute.

"Back in December of 2001, our apartment completely burned down and we lost everything. The company my father worked for ended up paying for a fully furnished apartment until ours was rebuilt," Selina said. At this point in time, we lived in Madison, Wisconsin. But that place burned down, so we were sent to [another place in town]."

---

43. Keel, *The Mothman Prophecies*, 78.

Then, in January of 2002, only two weeks after the fire, Selina and her mother, Heather, encountered the impossible.

They had just turned off their exit and were coming up on the second set of stoplights in the series of intersections, when, Selina said, her mother saw a "large batlike creature."

The creature was large and black, and when Heather turned to look at it, it turned and looked back at her.

Heather met the thing's gaze and saw that the eyes set in its batlike face were burning red as it flew by without flapping its wings.

The monster had been flying toward them and continued up and over their car before Heather lost sight of it.

"I was in the car when this happened," Selina said. "I didn't see what she did, but I do remember my mom screaming and panicking."

A few weeks later, when the family was able to move back into their original apartment, Heather saw the creature again.

She was in the basement doing laundry, directly across from which was the family's storage unit. At that moment, something compelled Heather to turn around. When she did, behind the accumulation of junk piled high in the unit she saw a large batlike face with those same red, glowing eyes, staring at her.

In a panic, Heather raced upstairs, so disturbed by her encounter that she never went into the basement again.

Shortly thereafter, the family moved out of the building, but those red eyes remain burned in her memory to this day.

However, that's not the only weird event experienced by the family at that apartment.

According to Selina, "Almost everyone in that building who was open to [the paranormal] really thought the building was haunted."

"I do remember a few instances where we would hear a baby crying. At the time, my brother was an infant, so my parents would go to check on him and he would be dead asleep, and they would still hear a baby crying. It was constant, it was almost nonstop," she said. "Some nights, they would hear all of the dishes flying out of the cabinets and crashing on the floor, and they would think that we had an intruder, and everything was untouched—it was like nothing even happened."

But, she said, "The activity really started getting worse when the basement started getting redone, because they were doing something in the basement. The activity started picking up after that. The apartment fire happened soon after."

This activity continued after the restoration with another concerning incident involving fire.

"I remember, soon after the building itself was rebuilt after the fire, my aunt and uncle had been living in the upstairs unit above us, and I remember being up in their apartment once and my aunt was braiding my hair, and I went downstairs because she was going to jump in the shower," Selina said. "They always had a decorative candle on their table—my family is Mexican, so they had those religious decorative candles—and they had one in the center of their coffee table. They didn't really light them up at that point in time, they never had a reason to, and so like I said, I went downstairs, my aunt was in the shower, my uncle was in the bedroom, and when my aunt got out of the shower, she noticed that the entire apartment was full of smoke. She thought, 'Oh my God, there's another fire.' It was the candle. The candle itself had set on fire—it wasn't lit, it had just set on fire. It was in the center of the table, and it even left a charred circle where it had been."

The intrusive and oddly threatening nature of the events experienced by Woodrow, the Lillys, and later Selina and Heather is oddly reminiscent of another aspect of reported goings-on surrounding the Mothman investigation that has since received a lot of attention—the Men in Black.

CHAPTER 4

THE MEN IN BLACK SUIT UP

Perhaps the first reported encounter with the Men in Black (MiB) happened when harbor patrolman Harold Dahl was approached by a mysterious man wearing a black suit following his encounter with several UFOs in July of 1947.

MiB, of course, is the moniker given to any mysterious figures who seem to show up out of nowhere following a strange event. They are known for asking odd questions and even making vague threats, all while keeping their identities thoroughly obfuscated. Just such an appearance was made by several anonymous people during the series of Mothman sightings in Point Pleasant, West Virginia. But before we delve too deeply into that, it is best we understand their origin.

As chronicled in Gray Barker's *They Knew Too Much About Flying Saucers*, Harold's experience began when he and his crew spotted "six huge doughnut-shaped objects in the sky" while patrolling at Maury Island near Tacoma, Washington.

According to the description given of the objects, they appeared to be one hundred feet in diameter and "of bright metallic coloring," with portholes spaced around their exterior. Inside the portholes were "dark, circular, continuous windows."

"Five of the objects were circling around the sixth, which seemed to be in mechanical trouble," Barker wrote.[44]

As the crewmen watched, the sixth object issued a muffled explosion, followed by the discharge of a "great quantity of metallic residue, something like lava rock, which fell all around them." Some of this debris reportedly struck the boat, causing considerable damage while killing a dog and injuring Harold's son. This explosion had seemingly obliterated the object, and the five remaining UFOs flew off.

Harold later told of his experience to fellow patrolman Fred Crissman, who was said to have helped collect sam-

---

44. Barker, *They Knew Too Much About Flying Saucers*, 148.

ples of the "slag-like residue" along with some "mysterious white metal" that accompanied it after the explosion.[45]

The day after his sighting of the odd aerial accident, Harold said he was approached at his home by a mysterious man in a black suit who invited him to breakfast.

The black-suited mystery man was reluctant to tell Harold exactly what it was he wanted to discuss as they drove to the restaurant, but once they were seated and ready to eat, he began to relate to Harold everything that had happened to the patrolman the previous day, "down to the most minute detail."

Harold was reportedly speechless, afterward saying that it was as though the man had been next to him that day, witnessing the same doughnut-shaped objects overhead.

As he sat there, stunned, the stranger began to threaten him.

"What I have said is proof to you that I know a great deal more about this experience of yours than you will want to believe," the MiB said ominously.[46]

He went on to say that Harold and Fred had seen things they shouldn't have, although he wouldn't elaborate on what exactly that meant. What the man was clear about, according to Harold, was that if he loved his family and

---

45. Barker, *They Knew Too Much About Flying Saucers*, 148.

46. Barker, *They Knew Too Much About Flying Saucers*, 150.

didn't want to see anything bad happen, then he wouldn't discuss his story with anyone.

This was all related by Harold to UFO witness and pilot-turned-investigator Kenneth Arnold, who had been sent by magazine editor and early supporter of weird stories Ray Palmer to investigate Dahl's claims in advance of publication.

According to *Project Saucer*, a digest of preliminary studies on flying saucers published in 1949 by the Air Material Command out of Wright Field in Dayton, Ohio, Arnold had from Tacoma "summoned two officers of Army A-2 Intelligence to aid in the investigation of Dahl and Chrissman's claim," named by Barker as Captain William L. Davidson and Lieutenant Frank M. Brown.

Harold is said to have met with Arnold, along with Captain Emil J. Smith, who had accompanied Arnold on his trip, and the two intelligence officers at the Winthrop Hotel, where he recounted events and provided them with samples of the fragments he said struck his boat.

Captain Davidson and Lieutenant Brown would take some of the fragments with them for technical analysis, planning to fly to Hamilton Field in California the next day to participate in an Air Force Day program. Tragically, they never arrived.

The plane they were flying in crashed, killing both men, although the crew chief and another passenger were able to parachute to safety.

Ultimately, the Air Force would deem Harold's story to be a hoax, citing a lack of credible evidence, although this likely comes as no surprise to anyone familiar with the Pentagon's history of publicly dismissing UFO sightings.

And while Harold's tale might represent the earliest—and very likely the deadliest—encounter involving the MiB, it is far from the strangest. For that, one must turn to Albert K. Bender, the man who truly ensconced the MiB within the annals of high strangeness.

Bender's journey into the unexplained begins with The International Flying Saucer Bureau (IFSB), an organization he founded in 1952 to investigate the mysterious flying saucers seemingly crowding our airways.

The IFSB was headquartered in Bender's home in Bridgeport, Connecticut, which he shared with his mother Ellen and stepfather Michael Ardolino. Bender lived in their attic and worked as chief timekeeper at Acme Shear Co., the largest scissor and shear manufacturer in the world, having been honorably discharged from the US Air Force in 1943 after serving as a stateside dental technician during World War II.

In an article for the Bridgeport Library's Bridgeport History Center, author Michael J. Bielawa described Bender as an exceptionally eccentric man who decorated his attic room—dubbed his Chamber of Horrors—with a variety of monstrous features, including shrunken heads, faux skulls, and even a spooky soundtrack featuring sobbing, hissing, and peals of thunder. To make his living space even odder, Bielawa wrote, "Bender filled his living space with an assortment of twenty chiming clocks." This ironic salute to his profession showcased Bender's humor, according to Bielawa, in addition to causing a cacophony of ringing bells at regular intervals.[47]

It was here, in his attic of oddities, that Bender first encountered the MiB.

As part of his investigation into the flying saucer mystery, Bender began publishing *Space Review*, a periodical associated with the IFSB. It featured a variety of articles on UFOs, many of which were eyewitness accounts, written by IFSB members—including Gray Barker, the IFSB's representative in West Virginia. And at some point during its publication, amidst the tales of saucer sightings and encounters with their otherworldly occupants, Bender claimed that he had discovered the secret truth behind it all.

This truth, it turns out, would come with a visit that so scared Bender he would stop publishing *Space Review* all together.

Bender had claimed to be experiencing a wide range of unusual phenomena prior to this, which he thought could be attributed to his interest in UFOs—his health had been poor, but even stranger, he said he had been receiv-

---

47. Bielawa, "Bridgeport's UFO Legacy: Men in Black and the Albert K. Bender Story."

ing bizarre phone calls and even telepathic communications warning him to quit the IFSB. He felt as though he was being watched, and even noted one night in November of 1952 that a man with glowing eyes seemed to be observing him at a local theater. Bender thought strange, shadowy figures were following him. Then, one night, he was reportedly accosted by a bluish flash while walking home.

"I was on a dark section of Broad Street when I suddenly developed a throbbing headache, and my ears seemed to block up," Bender wrote in his work *Flying Saucers and the Three Men*. "I felt as if something were pulled over my head to shut out everything about me. For some reason I looked up at the sky and, when I did, I saw a bluish flash. At the same time I had the sensation that my feet were being lifted off the ground. My head throbbed, and again, as when I had received the strange telephone call, I had the strong impression that somebody or something was telling me to forget IFSB, to give it up. As suddenly as the feeling came it left, and my head ceased to ache."[48]

When he returned home, Bender discovered that the blue light had followed him. In his den he found a "large object of undefinable outline" glowing in the center of the room.

"It looked like a bright, shimmering mirage," he wrote.[49]

---

48. Bender, *Flying Saucers and the Three Men*, 30.

49. Bender, *Flying Saucers and the Three Men*, 31.

Also present was the scent of burning sulfur, so strong it irritated his eyes. Sulfurous odors would continue to haunt Bender throughout his experiences. That specific smell is associated with a variety of paranormal phenomena, from Bigfoot to UFO occupants to demonic beings and other unexplainable things.

When Bender refused to capitulate to the telepathic demands that he quit the IFSB, he was visited by three strange men wearing all black.

"They looked like clergymen, but wore hats similar to Homburg style," he wrote. "The faces were not clearly discernible, for the hats partly hid and shaded them."[50]

The men, Bender said, were floating a foot off the floor. Suddenly, their eyes lit up "like flashlight bulbs" and focused on Bender, who felt all fear leave him as they did. The decidedly strange men in black then proceeded to deliver the telepathic message that prompted their visit. They acknowledged Bender's dedication to the UFO mystery but warned that his interest in the subject could do him harm.

It had been decided that Bender would be a very good contact for them on this planet, because he is an average person and thus unlikely to be believed if he told of what he'd experienced. Then the otherworldly visitors admitted that they were merely adopting a disguise to communicate with Bender, instead of appearing in their natural form.

---

50. Bender, *Flying Saucers and the Three Men*, 90.

"We have found it necessary to take on the look of your people while we are here," they told him. "This is mainly used as a means of returning here without being detected by anyone."

Apparently, the beings made regular contact with Earth by piloting craft from a secret base, but found it necessary to go to extremes to frighten off Earth people. Sometimes, they said, this had even resulted in the witness's death.

And as if what they were saying wasn't disturbing enough, they added, "We also found it necessary to carry off Earth people to use their bodies to disguise our own."

This baffling confession was necessary because at some point in the future they wanted Bender to write about them, and although nobody would believe him, he would be much wiser than anyone else on his planet. Not only would Bender come to know what was out there in space, they said, but he would also come to know the future of humankind.

Declining to give him their names, they insisted Bender simply refer to them as Numbers 1, 2, and 3. Before they left, the vexing entities gave Bender a small piece of metal, similar to a coin, that he was to keep in a secret place. The coin, when held, could be used to communicate with them.

"We wish to have you come with us at a time to be announced to you soon," they added.[51]

And then they were gone.

---

51. Bender, *Flying Saucers and the Three Men*, 91.

Eventually, the strange visitors would keep their promise, and Bender would report many more unusual occurrences. The paranormal harassment continued until finally, in October 1953, Bender released the final issue of *Space Review*.

It included the following: "The mystery of the flying saucers is no longer a mystery. The source is already known but any information about this is being withheld by orders from a higher source. We would like to print the full story in *Space Review* but because of the nature of the information we have been advised in the negative. We advise those engaged in saucer work to be very cautious."

Bender's experiences were subsequently published in the local news, striking a serious blow to his standing in the community and costing him many friends.

He went on to publish his autobiography, *Flying Saucers and the Three Men*, nearly a decade after *Space Review* was discontinued. Three years after that, he dropped out of ufology and moved to California, and on March 29, 2016, he passed away at the age of ninety-four.

The period spanning 1966 to 1967 during the height of Mothman mania along the Ohio River Valley saw a veritable invasion of odd characters intruding into the lives of witnesses. They came in slyly like something from a bad spy novel, their dialogue unconvincing and their motivations immediately suspect. There was Indrid Cold, of course, but he was far from alone. Cold was accompanied by a cast of nefarious strangers who seemingly defied conventional explanation; sometimes threatening, often confusing, and always mysterious, these unwelcome visitors were said by some to represent the same baffling phenomenon that had terrified Albert Bender into silence. The MiB and their ilk, it was said, had come to Mothman country.

Keel wrote in *The Mothman Prophecies* of a "blonde woman in her thirties, well-groomed, with a soft southern accent" who had reportedly visited some of the witnesses he'd interviewed, introducing herself as his secretary. He said that she carried a clipboard that "held a complicated form filled with personal questions about the witnesses' health, income, the type of cars they owned, their general family background, and some fairly sophisticated questions about their UFO sightings." These questions were proof, Keel surmised, that she could be no "run-of-the-mill UFO buff."

"I have no secretary," Keel wrote. "I didn't learn about this woman until months later when one of my friends in Ohio wrote to me and happened to mention, 'As I told your secretary when she was here ... ' Then I checked and found out she had visited many people, most of whom I had never mentioned in print. How had she located them?"[52]

52. Keel, *The Mothman Prophecies*, 55.

Keel's investigative partner, local journalist Mary Hyre, was one of those visited by a most unusual individual during this period. In January of 1967, Hyre was working late in her office across from the county courthouse in Athens, Ohio, when a small man entered. He stood four feet six inches tall, she estimated, and had dark eyes partially hidden behind thick glasses that were set deep beneath his long, black bowl cut. The little man's shoes possessed remarkably thick soles, which likely added an inch or two to his height. Although it was a frigid 20 degrees outside, he wore only a "short-sleeved blue shirt and blue trousers of thin-looking material."

He asked Hyre for directions to Welch, a town in southeastern West Virginia. His voice was low and halting, something the uneasy journalist interpreted at first as some sort of speech impediment. As he spoke, his unwavering gaze remained locked hypnotically on Hyre.

"He kept getting closer and closer," she told Keel. "His funny eyes staring at me almost hypnotically."

The odd man proceeded to tell her a long, winding tale about how his truck had broken down in Detroit, Michigan, and as a result, he had hitchhiked all the way to Ohio. He began to inch closer to Hyre as he spoke, and she became frightened. Thinking he might be unstable, she retreated to the back room where the *Athens Messenger*'s circulation manager was working, and together, they continued the conversation with the unsettling stranger.

"He seemed to know more about West Virginia than we did," Hyre would later say.

As their strange meeting continued, the telephone rang. When Hyre went to answer it, the diminutive man picked up a ballpoint pen from her desk and appeared to examine it with amazement as if, she said, "he had never seen a pen before."

Hyre offered him the pen and he responded with a "loud, peculiar laugh, a kind of cackle."

After that, he ran out of the office and disappeared around a corner.

The following day, she decided to check with the sheriff's office to see if any "mentally deficient person" had been reported in the community, but the answer was no.[53]

Months later, in the spring of that year, she would see him again.

Hyre's next encounter came in May, shortly after her husband and several neighbors said they had seen a luminous object appear over their house and project a "powerful beam of light" into their backyard.[54]

This time, when she saw him, he was on the street.

He had traded his blue short-sleeved shirt and trousers for a khaki uniform but wore the same platform shoes as before. When he noticed Hyre approaching him, he became

---

53. Keel, *The Mothman Prophecies*, 56.

54. Keel, *The Mothman Prophecies*, 107.

alarmed and ran off, jumping into a black car driven by a very large man.

"By the time I got out in the line of traffic he was gone across the bridge into Ohio," Hyre said. "I didn't get the license number, but the color looked orange."

Only three days later, on May 8, 1967, Hyre was arriving home at around 11:30 p.m. when a black car pulled up abruptly in front of her house. She stood there, on her porch, as a man holding a camera exited the vehicle and took her picture.

"His flashgun was very bright. It blinded me momentarily. While I was standing there rubbing my eyes he got back into his car, and it drove off. I couldn't see if there was anyone else in the car," she said to Keel before asking, "Now why do you suppose anyone would want to take my picture like that?"[55]

Another witness would report a similar encounter, explaining that it occurred months after being chased by a winged creature along Route 2 "near the Chief Cornstalk Hunting Grounds."

The man's ordeal began in April when, one rainy night, a large black figure emerged from the woods and flew over his car.

"It was at least ten feet wide," he told Keel. "I stepped on the gas, and it kept right up with me. We were doing

55. Keel, *The Mothman Prophecies*, 107.

over seventy. It scared the hell out of me. Then I saw it move ahead of me and turn toward the river."

Later, in October, after returning home from work one night, the man opened his door to find an unexpected visitor.

"When I opened the door, I saw this man standing in my living room," he said. "I think he was dressed all in black. I couldn't see his face, but he was about five feet nine. I started to fumble for the light switch when he took my picture. There was a big flash of light, so bright I couldn't see a thing. While I was rubbing my eyes the burglar darted past me and went out the open door. I guess I arrived just in time because nothing was missing."[56]

As disturbing as these invasions of privacy are, they pale in comparison to the danger represented by other witnesses' experiences with the MiB.

Less than a month after her November 16, 1966, UFO sighting over Point Pleasant's TNT area and subsequent Mothman encounter, Marcella Bennett narrowly escaped the MiB.

In early December of that year, Marcella was driving with her infant daughter, Tina, on one of the deserted back roads outside of Point Pleasant when she noticed a red Ford Galaxie following her. In the driver's seat was a stranger—

56. Keel, *The Mothman Prophecies*, 107.

a large man who appeared to be wearing a "bushy fright wig."[57]

Marcella slowed down for him to pass, but he chose instead to overtake her and try to force her car off the road. The frightened mother hit the accelerator and the Ford sped past her until she lost sight of it around a curve. Rounding the bend, Marcella was startled to find the other car waiting in ambush. The man had parked it across the road in an apparent attempt to stop her. Now terrified, she slammed her foot on the gas pedal. Once the apparent carjacker saw that she wasn't going to stop, he quickly pulled his car out of the way to avoid a collision.

Marcella had never seen the man before and she never saw him again, but much like the other harrowing events she'd experienced so far, he stayed with her all the same.

But perhaps the most alarming story is that of Mary Hyre's niece, Connie Carpenter.

On Sunday, November 27, 1966, Connie was driving home from church at around 10:30 a.m. when, as she passed the vacant Mason County Golf Course outside of New Haven, West Virginia, she ran headlong into the impossible. Standing on the course's well-manicured lawn was a huge, gray, humanoid figure. Connie described the figure as "at least seven feet tall and very broad," but what really got her attention was its eyes. The thing's eyes, she said, were large,

57. Keel, *The Mothman Prophecies*, 55.

round, and glowing red with a fierceness that locked her into a hypnotic trance.

"It's a wonder I didn't run off the road and have a wreck," she later told Keel.[58]

She slowed, her eyes never leaving the creature, watching as a pair of wings unfolded from its back. She guessed their span at about ten feet. As she watched, the being rose silently, straight up into the air, its wings outstretched but otherwise motionless. In a panic, Connie slammed the accelerator down as the creature swooped low overhead, escaping its strange hypnotic gaze.

Afterward, Connie's eyes became red and puffy with a case of conjunctivitis that would last two weeks, and she claimed to occasionally hear loud beeping sounds outside of her bedroom window.

Sometime later, a white car appeared at Connie's home in New Haven, West Virginia. It reportedly sounded like it had a faulty muffler and announced its presence with a loud clunking as it pulled up. Out of this seemingly mundane vehicle came a man who introduced himself as Jack Brown.

Jack proceeded to ask a series of unusual and off-putting questions to Connie, her brother Larry, and soon-to-be husband Keith Gordon, his strange demeanor making all three unwilling participants uncomfortable. They said he wouldn't speak unless they looked directly into his "dark, hypnotic

---

58. Keel, *The Mothman Prophecies*, 13.

eyes." And more things than just his eyes were unusual. The trio noted his curiously long fingers and thought that there was "something very peculiar about his ears," although what exactly that was, they couldn't quite place.

"I'm a—a friend of Mary Hyre's," he began.

The awkward fellow didn't seem interested in Connie's Mothman sighting, but instead focused on Hyre and her relationship with Keel.

"What do you think—if—what would Mary Hyre do—if someone told her to stop writing about UFOs?" he asked.

To which Connie replied, "She'd probably tell them to drop dead."[59]

He continued this line of inane questioning for some time before returning to his car and noisily rattling into the night.

Connie called her aunt immediately after the strange visit, confused and upset by the man's odd behavior, but there were no answers to be had.

In early February of 1967, Connie and Keith were married and moved across the river to a house in Middleport, Ohio. Middleport was a small town of roughly three thousand people, and only a handful of family members and close friends knew their new address, nor did they yet have a phone.

---

59. Keel, *The Mothman Prophecies*, 14.

Connie was still a student at that time, and so it was at approximately 8:15 a.m. on February 22 that she set out to walk to school after Keith had already left for work.

The quaint, tree-lined street was quiet, and as she walked, a remarkably well-kept black 1949 Buick pulled up alongside her. Sitting in the car's spotless interior was a "clean-cut young man of about twenty-five" who was "wearing a colorful mod shirt, no jacket, despite the cold weather." He had black hair that was neatly combed, a deep suntan, and no noticeable accent as he opened the door to ask Connie for directions.

She walked helpfully over to the stranger's car, but as she arrived, he lunged at her, grabbing her arm and ordering her into the car. Connie fought back and managed to break his grip, running back to her house. Terrified, she locked the door and didn't leave until Keith came home from work. Still shaken up, she stayed home the next day too.

At around 3:00 p.m., Connie heard someone on the porch, followed by a knock on the door. With an abundance of caution, she waited for some time before checking the door. She found the porch empty but noticed someone had slipped a note under the door. Written in pencil in block letters on an otherwise ordinary piece of notebook paper, it read, "Be careful girl. I can get you yet."[60]

---

60. Keel, *The Mothman Prophecies*, 63.

Connie and Keith went to the police that night to report what had happened, turning the note over to officer Raymond Manly. By March of 1967, when John Keel arrived in the hope of examining the note, it had disappeared. Officer Manly claimed it must have been lost, which might come as no surprise, given that the entire file detailing the frightened young woman's experience amounted only to "a printed form containing Connie's name and address and one scribbled line, 'Dark Buick, young man.'"

According to Keel, "The police chief assured me that no such car existed in Middleport and that it was obviously a case of some maniac trying to abduct a young girl. Officer Manly told me he was keeping the house under constant surveillance. I had to break the news that the Gordons had moved back to the West Virginia side of the river shortly after the incident."[61]

Connie had escaped the clutches of high strangeness, but not without bearing the emotional scars so commonly held by those lucky enough to do so.

And yet all the fear and apprehension caused by these mysterious, threatening interlopers amounted to so very little next to the awful, inescapable tragedy that would eventually come to Point Pleasant.

---

61. Keel, *The Mothman Prophecies*, 63.

CHAPTER 5
THE LAKE MICHIGAN MOTHMAN

Just over fifty years after the creature was spotted in West Virginia, Mothman arrived in the Midwest.

It began with a trio of sightings in the spring of 2017 reported to the Mutual UFO Network (MUFON). Three separate witnesses reportedly saw an unidentified winged humanoid over Chicago between 10:00 p.m. April 15 and 2:00 a.m. April 16.

LAKE MICHIGAN SIGHTINGS

The first case, numbered 83325, was reported by a woman who was out on Lake Michigan with her husband and two other couples celebrating a friend's birthday.

According to her report, the group was "about two miles out on the lake, just off of Montrose at about 10:00 p.m." when their encounter occurred.

"We were enjoying ourselves when I happen to look up and saw what looked like a giant bat, and not like a fox bat (which I looked up and saw was the biggest bat)," she said. "This bat was as tall as my husband, who is six foot four inches [tall], or even bigger."

She described the creature as "solid black with eyes that seemed to reflect the moonlight." The enormous bat was "blacker than the surrounding night sky" against which it was "perfectly silhouetted" as it circled their boat in complete silence. After flying around the stunned partygoers three times, the monster left in the direction of Montrose, where it "quickly blended into the night sky and was gone in seconds."

Also, according to the woman's report, a bright green object traveling north to south at the horizon was seen about five minutes after the creature flew off.

"It was not a plane as it was brilliant green and was moving slowly across the horizon," she added. "If I had to estimate, it was about two miles from our position. After the object was out of sight, we sat there looking around in stunned silence. I began feeling this overwhelming sensation of dread and told my husband that I felt it was prudent we get off the water as quickly as possible. I tried to get a picture of the thing as it circled our boat, but all I got was black."

Weather conditions at the time mostly matched the woman's version of events—the moon was waning gibbous, and the sky was partly cloudy—except moonrise that night was at 11:45 p.m. So, either the witness was incorrect about the timing of the event, the detail regarding the reflection of moonlight was embellished, or she mistook a different light source for the moon.

The second report, Case 83243, took place thirty minutes later, at 10:30 p.m. This witness was said to be "hanging out with my boys and a few friends" in Chicago at the time of their sighting.

"As we talked about work and our families, we heard what sounded like a bird flapping its wings," the witness wrote in their report. "One of my homies yelled out that he saw a huge Lechuza over by the road."

In Mexican folklore, a Lechuza is an old woman—often a witch, or *bruja*—who can turn into a giant blackbird. In most stories, the bird is an owl, but sometimes it is described as an eagle.

"We walked over there and saw what looked like a big owl," they continued. "As we walked up on it, this owl stood up on two feet and looked right at us. We saw what looked like a huge Lechuza, except it was about six feet tall and really big. It had large glowing red eyes that were completely freaking everybody out. We all yelled, and this thing took off into the air and took off toward North Avenue. This thing freaked us all out and scared all our kids."

The third report, Case 83206, occurred at 2:00 a.m. on April 16, 2017.

This witness claimed that they had just arrived for work at the Chicago International Produce Market when, as they crossed the parking lot, they noticed "four or five guys" staring up at the sky. Naturally curious, they investigated.

"I looked up and saw the biggest freaking owl I have ever seen! I'm six foot two inches [tall], and I'm guessing this thing was at least a foot taller than me," their report read. "It was completely black except for it having bright yellowish/reddish eyes like a cat."

Whatever it was, it made everyone present uneasy, they said.

The creature proceeded to stand there for a minute or two before "shooting up into the sky and disappearing" after some people watching it began to throw rocks.

"It had wings on it like an owl, only bigger, and you could hear it flap those wings when it took off," the witness said. "It made this sound as it took off and flew away that sounded like a truck's brakes when they are burned out."

The owlish monster "flew up and flew a wide circle, making that sound once more and then flew off in the direction of the Stevenson Expressway."

After the bizarre being had left, the group of onlookers spent several minutes continuing to look for it, but it never came back. This creature, whatever it was, had made an impact.

"I'm reporting this because there is no way this was an owl," the witness explained. "It stood upright like a man, just really, really tall."

Furthermore, they said, "I don't want to discuss this with anyone and would prefer to remain anonymous. I don't imagine anyone would believe me anyways. One of the guys I work with who saw this thing said that it made him feel very uneasy, like a scared kid, and he was glad someone threw a rock and made it fly away."

In an article published on their website, MUFON compared the sightings to those of West Virginia's Mothman.

MUFON advised that "Illinois MUFON State Director Sam Maranto and other staff members are investigating. This account is taken merely from the early witness reports filed with MUFON. Please remain skeptical until witness interviews have been conducted and more case evidence is assembled."

However, Maranto later admitted to The Singular Fortean Society that only limited email communication was ever had with the witnesses, and that two of three IP addresses in the emails were found to match. Despite the matching IP addresses, it is uncertain whether the emails were sent from the same physical address, since many of the large email companies—including Google and Microsoft—only allow the IP addresses in their emails to be traced as far as the last company server that they passed through. Assuming that the emails were written by unrelated persons and simply passed through the same server, this explains the matching IP addresses without there having to be any deliberate hoaxing involved.

Which leaves us with a mystery; one that probably would have been quickly forgotten had these reports been the only clues present. But that's not what happened. No, these reports kickstarted a slew of winged creature sightings that dwarfed anything seen before. For better or worse, the Mothman was here to stay.

LITTLE VILLAGE SIGHTINGS

In 2017 alone, over fifty sightings were reported. Rather than flooding into MUFON, these new sightings were largely being reported to two websites: UFO Clearinghouse and Phantoms & Monsters, owned and operated by Manuel Navarette and Lon Strickler, respectively. Shortly after this deluge of winged humanoid reports began, The Singular Fortean Society reached out separately to both men for comment.

Navarette, a native Chicagoan, said at the time that he had been canvassing neighborhoods where sightings were taking place, such as Little Village.

Most sightings from Little Village were reported to Navarette through his website UFO Clearinghouse. According to Navarette, residents of the largely Mexican American neighborhood would report terrifying sightings of a winged monster they would refer to as a Duende (Spanish for "goblin") or La Lechuza, the witch referenced in one of the earliest reports to MUFON. Most of these reports would go unverified by other investigators, an unfortunate trend seen throughout the first year of Mothman sightings reported in the region.[62]

Strickler, a resident of Pennsylvania, was unable to conduct onsite investigations, although he received dozens of reports through Phantoms & Monsters.

Despite his distance, Strickler was certain that something strange was happening in Chicago.

"These reports have been very specific," he told The Singular Fortean Society in a 2017 interview. "Even when I prodded them to embellish, they wouldn't. They haven't embellished. I don't think people are imagining it, and they come from all different ranges of lifestyles."

Soon enough, The Singular Fortean Society's coverage of the sightings meant that they began to receive reports directly from witnesses, and by 2018 they had officially joined the investigation.

As the investigation grew, so did the area from which reports were received.

---

62. Wayland, "Flying Humanoid Fever Hits the Windy City, an Interview with Manuel Navarette."

PRAIRIE CREEK RESERVOIR SIGHTINGS

One hotspot to emerge from this expansion was the Prairie Creek Reservoir in Indiana, about five miles southeast of Muncie and 140 miles from Lake Michigan.

The park's defining feature, of course, is the reservoir itself, which amounts to more than 1,200 acres of water for folks to enjoy, surrounded by an additional 750 acres of rustic beauty, and it offers everything from boating and swimming to hiking and horseback riding. It's a popular recreational area for a reason, and there are usually plenty of people around to witness anything unusual should the opportunity arise. And arise it did.

Strickler received his first sighting report from Prairie Creek Reservoir in early 2019 and spoke to the witness, who asked to remain anonymous, via telephone soon after.

The sighting took place on December 26, 2018, between 4:00 and 4:30 p.m.

"[The witness] and his wife were traveling southbound on a county road, about a mile south of the Prairie Creek Reservoir near Muncie, Indiana. The date was December 26, 2018, at dusk. A huge flying object caught [his] attention," Strickler wrote in his blog. "[The witness] is a military veteran, hunter, trapper and farmer who lives in the immediate area. His knowledge of military flying craft, wildlife, and his keen sense of observance was apparent while I talked to him. The winged being that he was observing was unlike anything he had ever seen before. The creature was flying just above treetop level and was easily visible to the witness. His reaction was to slam on his brakes in wonderment, exclaiming to his wife, 'Do you see that?' His wife was shaken by the sudden stop and was unable to react fast enough to see the winged being."

The creature described by this witness bore a similarity to the reports of winged humanoids nearer Chicago, including the stillness of its wings in flight.

"[He] stated that the being was humanoid in shape with an obvious 'face,'" Strickler said. "The body had a length of approximately six to seven feet, with batlike wings that were extremely wide. The being was dark colored and seemed to glide at a steady speed. He never noticed the flapping of wings while watching the creature."

According to the witness, the creature's wings had a leathery texture. He was unable to discern if the creature had any additional limbs beyond its wings, nor could he recognize fine detail in its face.

Also similar to the other sighting reports was the profound effect the experience had on the man.

"[The witness's] wife states that [he] has been truly affected by the incident and has constantly mentioned it to her, in an attempt to explain what this winged being really was," Strickler said. "He had refused to mention the incident outside of his family. When [he] read about the recent sightings in Gary, Indiana, he called his wife from his job and asked her to contact me right away. He later called me when he got home. This witness was very forthcoming and anxious to find out what this creature was."

This report was soon followed by another sent to Strickler from someone who said it reminded her of her own experience. This witness also elected to remain anonymous.

Her encounter, she said, took place in the winter of 2011.

This witness submitted her report via email and declined to be interviewed over the phone, citing privacy concerns, but Strickler insisted at the time that she "seems very honest" and proceeded with publishing her sighting to his blog.

According to the woman, her husband was on Facebook when he spotted an article about the previous report and showed it to her.

"I am certain this is the creature I saw eight years ago," she wrote.

She was driving in Randolph County, just "ten to fifteen minutes kind of east from Prairie Creek Reservoir," when she "saw what appeared to be a human crouched down in the road."

However, as she approached the figure and slowed down, eventually coming to a stop as it refused to move, it turned to look at her with greenish-yellowish eyes, "like the color of cat eyes."

"It stared for maybe three seconds and then proceeded to slowly get up from its position, like a human would stand up because it had legs," she said.

The thing then turned toward her car, stepping on the hood and launching itself into the air as it "lifted off like a glider toward the sky."

"I turned my head just in time to see it tuck its legs into its body, so it did appear as a bird in flight... kind of," she explained. "Its wingspan was wider than anything I've ever seen on an animal in my life."

She told her husband of the experience, and he expressed his opinion that she had simply seen a bird; an act that caused the witness not to speak of the encounter again in fear she would "be made out to be an idiot," at least until she saw that article on social media eight years later.

This causal chain of experience recognition added a new link less than a month later when another witness came forward. Her mother had sent her one of the previous articles because of something the witness experienced back in the summer of 2007.

This witness also asked for anonymity, this time due to being a "known credible media professional," Strickler said. The Singular Fortean Society would later verify that claim.

She explained she was originally from Muncie and spent a great deal of time at Prairie Creek Reservoir because her family has a boat that they take out on the water there.

It was dusk as she and her friend headed toward shore aboard the family boat. They had been out enjoying the sunset and were returning to the dock when they noticed a strange man standing at the end of it. There was no one else around; all her neighbors' boats were docked, no cars were parked nearby, it was just them and the strange man. He turned as they approached to reveal a startling set of "glowing, yellowish eyes."

"At this point, we were freaked out wondering what this was because there was something totally off about it and it was clearly not a normal human," she said. "Then, as we got even closer, it spread its wings, flapped a few times, and soared up into the sky."

As she took in the incredible sight, she decided that whatever this was it was too big to be a bird. The thing stood at least five or six feet tall and was very dark; its entire body and wings were grayish black in color.

"I've never seen anything like it before or after," she said. But this wasn't the limit of her unusual experiences at the reservoir.

"Not sure if this is of interest, but my family and I also saw a UFO at Prairie Creek Reservoir around dusk a few years later," she added. "We were one of the only boats on the water; it was such a serene, quiet, and peaceful evening. My mother looked up at the sky and saw the UFO, so we all looked up. It was shaped like a tin can, stayed still, and made absolutely no noise. At that instant, my father caught a fish, and the commotion caused us all to look down, and a few seconds later when we looked back up, it was gone without a trace."[63]

If these had been the only reports out of this part of Indiana, then perhaps they would have quietly passed into obscurity and been forgotten. But that's not what happened; instead, the anomalies kept on coming.

In 2019, Garry Patterson contacted The Singular Fortean Society to tell of how he had witnessed "many balls of light" in a rural area of Indiana near Prairie Creek Reservoir.

"I'm about seven miles from [the Prairie Creek Reservoir]," Garry said. "I've seen many balls of light in this area, from Farmland toward Muncie. They are intelligent. I've seen them react ... change color, move, or shut off."

The then-fifty-one-year-old man stated that his family moved to the area in 1979, and his first sighting of a glowing orb in the area was in the summer of 1981, when he was twelve.

"It was a clear summer night sky," he said. "I saw an orange/amber ball of light about twenty feet above the trees. It started toward me slowly, so I ran in and got my mom. When we came back out it was over our trees. The actual size is hard to say but I'd guess three feet across. It

---

63. Wayland, "Flying Humanoids over Chicago, an Interview with Lon Strickler."

was totally silent, and it moved off to the west out of sight behind the trees."

Garry couldn't help but wonder if, somehow, he might have initiated contact with the anomalous light.

"Funny thing, probably the summer before, I used to blink one of those old flashlights with the switch and button into the sky," he said.

Since his initial sighting in 1981, Garry said he's had a number of sightings of strange glowing orbs in the area.

"I'm guessing I've seen them eight or so times where they were either red or orange, and maybe three to four where they were white," he said. "The last time was a few years back [in the summer of 2016]. I was driving east coming home at night, when I spotted an orb that was orange, and it seemed like a second one right next to it was changing back and forth from red to orange and dimming to dark and back. So, I got to my road turning south and I came to a stop to watch. The car window was down, so I fumbled with my phone to hit record … snap, gone."

These sightings occurred in the area south of and between Parker City and Farmland, near Windsor, and align themselves with a two-thousand-year-old Native American burial plot in the shape of an oval that covers an acre of land. To the west of the sightings and the mound, in a nearly straight line, lie the Prairie Creek Reservoir and Mounds State Park.

The line formed by these sightings and the associated landmarks is less than two hundred miles north of the 37th Parallel, known as the United States' UFO Highway. Stretching from Santa Cruz, California, to the Chesapeake Bay, this area is known for UFO sightings, cattle mutilations, and other anomalous phenomena.

Some researchers suspect the 37th Parallel to run along a powerful ley line—paths drawn between landmarks that some believe represent a natural energy unrecognized by modern science.

"I know things are found along ley lines," Garry explained. "Since the Mothman sighting at Prairie Creek, that makes a rough line from there to here and also Mounds State Park."

As far as what he believes the orbs might be, Garry, who identifies as a Christian, could say only that he does not believe they are any technology known to man.

"I know what the various things at night look like," he said. "As to what I think is behind the orbs ... I believe they are part of the spiritual or supernatural realm. [They] do things aircraft or pilots could never do."

Following this conversation, Garry reached out again in 2020 to report another sighting of anomalous glowing orbs.

He was driving east, just south of Muncie, at approximately 12:40 a.m. on May 29 when he spotted three amber orbs.

The first orb he noticed was directly to his east, and soon after another appeared to the south.

Garry described the orbs as amber in color, and said there was no movement or flashing. They appeared coin-sized from his perspective, he said, but he couldn't discern what their actual size was.

"It's really hard to tell with a light, but most have appeared the same size and are usually the same color: amber/orange," he said. "I do know the one to the north looked bigger; it was a steeper angle, more above me. To be close enough or large enough for one to look larger above me, I would guesstimate three to four feet across, no smaller."

As he watched, "The one straight to the east shut off, so I looked around and there was one that seemed closer to the north."

He continued. "As I was driving, there were trees in the way on some parts. Then the one to the north shut off. I could still see the one to the southeast, then it was gone also."

Garry couldn't find any mundane explanation for the orbs.

"I saw no airplanes and it was cloudy to the east, with broken cloud cover overhead," he said. Historical weather data confirmed that the sky was mostly cloudy at the time of his sighting.

What exactly Garry has been seeing flying around over Indiana for the past several decades remains a mystery.

Garry wasn't the only one seeing strange things other than Mothman around Prairie Creek Reservoir.

A month after Garry had first contacted The Singular Fortean Society in 2019, they received a report from another man, wishing to remain anonymous, who said he and a friend had seen a "shadowy man" transform into "a big black shadowy catlike creature" after suddenly appearing out of a "flash of bright light."

According to the witness's initial report: "The following sounds crazy, but I promise it happens exactly as described. My wonder is if anyone has ever reported something similar. Years ago, a friend, myself, and my collie were sitting in my driveway around 11:00 p.m. just BSing on a summer night. Suddenly my dog began to go crazy and beg to go inside.

"Then, across the street, there was a quick flash of bright light, and a shadowy man stepped out of the light. The man turned to look at us, and as he / it did, it went down on all fours and went from a shadowy man to a big black shadowy catlike creature. There was another flash of light, and it was gone.

"During the event neither one of us spoke, but afterward I looked at him and said, 'Do you want to go inside,' he said, 'Yes,' and so we did. Inside we took turns describing the event to each other and both of us saw the exact same thing.

"Years later I ran into this friend again and he still remembers every detail exactly the same. No drugs or alcohol were involved, nor had either of us tried any of that yet. It was clear as day and I remember every detail like it was yesterday. I know we saw something; I am wondering if anyone knows what it was."

In subsequent correspondence, the man described a profound feeling of nothingness prior to the experience.

"It was very still, like before a tornado hits," he said. "The only one affected prior to it appearing was my collie, Rocky. He went nuts running from us to the side door, begging to go inside. Rocky was older and never acted that way. He would sometimes stay out all day or night just doing his thing. After it happens and we went in, that dog stayed next to us or at our feet and never budged.

"A light breeze was blowing but had stopped prior to [the event]. No crickets, cicadas, when I say nothing, I mean *nothing*. It was as if everything else knew something was up."

This disturbing encounter took place in Richmond, Indiana, at around 11:00 p.m., in either the third or fourth week of June 1993; he was not sure of the exact day.

Richmond is about thirty-five miles southeast of Prairie Creek Reservoir.

The sightings around Prairie Creek Reservoir led to other reports that were, unfortunately, left unverified. As is often the case, the witnesses who initially submitted these reports never responded to attempts to follow up. That doesn't mean they didn't happen, but it does have a negative impact on their overall credibility.

These unverified reports included a supposed encounter with a "a large, dark, winged creature" that flew up from a drainage ditch and over the car of one surprised motorist outside of Roanoke in 2019, another motorist who claimed they encountered a terrifying batlike creature crouched by the side of the road outside of Muncie that same year, and again in 2019 a passenger in their roommate's car said they saw a flying person keeping pace with them on Indiana State Road 3. Prior to those sightings, at least chronologically by when they purportedly took place, came the 2016 report from someone who said they were driving home from Muncie when they saw a creature startlingly similar to our previously mentioned batlike flying humanoid with "an obvious face."

WISCONSIN SIGHTINGS

Wisconsin had its share of sightings too. The southeastern part of the state is known for high strangeness; everything from Bigfoot to Dogman to UFOs to Mothman has been reported throughout the rural municipalities that bridge the gap between Milwaukee and Chicago. It was within this weird expanse that a man said he had encountered a humanoid "bat-dragon" in April of 2017, just before events kicked off in Chicago with those first three reports to MUFON.

Like so many others, he asked to remain anonymous.

The man had arrived at his home, located between Eagle-Palmyra and Mukwonago, sometime around 10:45 p.m. and decided to sit in his van and finish the conversation he was having with his friend rather than take it inside and risk waking up his parents (with whom he lived). Moisture hung heavy in the air; a continuation of the dreary drizzle that had continued all day and into the night. As he continued chatting with his friend, the man looked up and noticed something peculiar; silhouetted against the light shed by a lamppost fifty feet away at the end of the driveway—the only source of illumination—stood a "very, very tall" and "very dark" figure, not more than five to ten feet from the van's hood.

It took the man's brain a second to register what he was seeing, but when it did, he began relating to his friend what was happening as he turned on the headlights.

"I gave a concise yet detailed-as-possible description to my buddy while I was literally shaking and scared out of my wits," he said in his report. "I was looking at a creature, for lack of a better term, that essentially looked like a seven-foot bat/reptile of some sort. The head was at the level of the roof of my minivan, or slightly more elevated than my roof, and it was standing perfectly still, just staring right at me. Its eyes were large, taking up a significant portion of the thing's head, from what I could make out, and although they were dark, like large black eyeballs, there was a glint of reflection in them that allowed me to discern that they were, in fact, definitely its eyes. It was haunting, for lack of a better way to put it."

The thing possessed an "extraordinarily swarthy coloration, though slightly reflective, like skin or scales of some kind, not feather nor fur covered," and besides its height was also quite large.

"It was at least as wide as a very large man, but did not seem 'stocky,' as it were," he said, "and as I stared at it uncomfortably, I realized that it actually had huge wings, but they were wrapped around its body, exactly as a bat wraps its wings around its body while sleeping upside down."

Having difficulty processing the impossibility of what stood before him, the man blurted out to his friend that he was seeing a "giant bat-dragon-looking thing." His friend asked him if he'd been drinking, but the man insisted he was sober and that, yes, whatever this thing was, it was right in front of him. And then, just as he was done describing it, the monster vanished.

It seemed to move very suddenly, he said, without unfurling its wings, and in a blur, it was gone.

He largely kept the sighting to himself after that out of fear of ridicule, until other sightings started being reported.

"Even more unsettling," the man added, "was that about a week to two weeks later, that same old buddy I had been speaking with the night of the experience called me to inquire if I had watched or heard the news the previous evening. I informed him that I had not. After having initially assumed that I had gone completely out of my mind, for well over a week, he informed me that my sanity had been vindicated. Numerous encounters with the same creature had been recently reported in Chicago and northern Illinois, in largely populated areas."

And that was the last the man saw of his bat-dragon, whatever it was.

The man's testimony would remain consistent in conversations with both Lon Strickler and The Singular Fortean Society. He even invited Fortean members to his home to investigate.

During that visit, the man added that he had heard a "mixture of high-pitched screeching combined with screaming" several times outside of his home at night and speculated that perhaps it could be related. He also said that he spoke to his friend, and the only detail his friend would add to the incident is that the witness was silent for several seconds prior to saying anything about the encounter over the phone—presumably due to the shock of what he was witnessing.

About sixty miles or so west of Mukwonago is Oregon, Wisconsin, where two young women would have their own encounter with the impossible on Thanksgiving night in 2020.

This report first came to Jesse Durdel of the National Cryptid Society, who put The Singular Fortean Society in contact with the young women.

Both witnesses asked for anonymity. In the spirit of honoring that request but also keeping events clear to the reader, they have been provided with pseudonyms.

Claire and Ashley, both of whom are eighteen-year-old women from Oregon, Wisconsin, said in a phone interview with The Singular Fortean Society that they'd seen an eight-to-nine-foot-tall creature with a wingspan of ten to fifteen feet while driving outside of town on November 26, 2020.

Ashley said that she had initially seen something unusual while driving between Oregon and Stoughton on Highway 138 at around 10:00 p.m. that evening.

"I was on that road because I was going to Stoughton for some stuff," she said. "I was driving my car. I noticed this thing—I couldn't tell how big it was, it just looked like a black shadow—just start across the road, but it was probably a football field ahead of me. At first, I thought it was my windshield, because I have cracks in my windshield, but that doesn't make sense because I've driven that road hundreds of times and I've never seen my windshield do anything like that. It was insanely fast. I was like, 'I think I just saw something, but I'm not going to say anything because I'll sound crazy.'"

Later, at around 10:30 p.m., both women were going for a drive together when they said they saw a flying humanoid swoop across the road in front of them at the intersection of Sand Hill Road and Rutland Dunn Town Line Road.

"There was a possum in the road, and we stopped because we didn't want to hit it. We were driving around it, and then we looked up, and it was huge, like nine feet long," Claire said. "It was flying, and it swooped down across the road, and out of our sight. We were both like, 'You saw that too, right?'"

"[The creature] was like a quarter of a football field ahead of us. It was really close. It was way too big to be a deer," Ashley continued. "It was kind of reddish brown in color, I want to say, but it was darker. If we hadn't stopped for that possum, it would have hit our car. That thing was huge. My first thought was, 'Dear Lord, it's a pterodactyl.'"

A lamppost positioned at the intersection partially illuminated the being from underneath, they said, allowing them to determine its approximate color and dimensions.

According to the two women, the creature was headed north.

"It was too high up to be illuminated by the car's headlights. The very bottom was illuminated, but we couldn't see the top of it. It wasn't fully illuminated, but light was cast on the underside," Claire said. "Since it was at the intersection that we normally turned at—we normally turn at that intersection because we make loops around town—it swooped from the south toward the north. When you get to the intersection you have to turn north, and I was terrified to turn north. I didn't want to turn north."

Claire described the creature as looking "like Mothman" and said, "It looked like a very bulky person. Its limbs were a lot thicker than a human's [limbs] would be—a lot bigger around. It had these absolutely huge wings. There were arms on its side; actually, I didn't know if they were arms, there were appendages on its sides, and the wings were on its shoulders or back, but I couldn't really tell. I didn't see its head when it swooped down in front of us. I didn't see any head or neck when it swooped down."

They decided to continue their drive, eventually working up the courage to revisit the location of their sighting. At around 11:00 p.m., as they drove by the area again, Claire spotted something unusual in a nearby field.

Although the creature had first been seen heading north, when they returned, "It was standing on the south side [in a plowed cornfield]."

"We were at that point where we thought it would be gone, but then it wasn't," Ashley said.

Ashley, who was driving, didn't see the creature standing in the field. Claire, however, said she saw a strange creature with "very pronounced red eyes."

"I looked out the back window and [the creature] was standing in the field. It was almost a full moon, so it was pretty illuminated. Its head didn't look like it had a neck, it looked like it was just an extension of its body. It looked like a human had tucked their head into their shoulders.

"It had very pronounced red eyes. We didn't have any headlights in the back, and the only thing that would have reflected would have been the lamppost, so they were either reflecting really brightly or literally glowing. We go by that field all the time and I've never seen anything like that before or since."

The women described the creature as superfast, especially for its size.

"It was very fast, and the wings were very, very large," Claire said. "The body was so stocky. The silhouette was similar to how I would describe Bigfoot, which I know, I'm using other known cryptids to describe this one, but that's really what it looked like. There was no pronounced neck, but it looked very stocky. It had very large, birdlike wings. I couldn't tell if it had feathers."

Ashley said that she thought "the part illuminated by my headlights looked like reddish brown fur."

"It was either very fine feathers or fur," Claire agreed. "I didn't really see a texture. There weren't large feathers on the body—that I would have been able to see. The only thing I can say, is that when I saw the silhouette of it in the field, I could see jutting off points from the wings that I would call feathers, but I don't really know. It could just have wings shaped like that."

Both women said that they screamed after seeing the creature fly over the road, and Claire added that she froze with fear in response to the sighting.

"[I felt] fear, yeah, there was also the element of 'that's not supposed to be here,'" Ashley said. "You just know that whatever that is, it's not from around here. It's not supposed to be here."

"Yeah, like wrongness," Claire added. "We passed that place a couple of more times, and the rest of the night we just got a weird *there's something that shouldn't be here* feeling every time we were in that area."

They both also noticed other unusual phenomena happening that same evening, such as deer moving away from the spot where they had seen the creature and unusual interference on the car's radio.

"This was the weirdest thing that I noticed, it was way earlier, but it counts for the whole night basically. We kept seeing deer crossing the road," Ashley said. "I almost ran into two different deer on two different sides of town. Usually, if I see a deer at all, it's one deer a night. Oregon does not get a lot of deer, and I saw over seven deer that night. And we saw a whole herd running away from that spot."

"They were freaked out; they were running away. They were panicky," Claire chimed in.

The radio, Claire said, "kept glitching like crazy, and it hasn't really done that before. It was cutting in and out, getting staticky, and then picking up another channel that was some gospel program. I know what channel that is, and it's very far away from the channel we were on."

She added that the radio would glitch "only when we drove by that spot."

Only a few days later, on the evening of November 29, Claire said she had another unusual sighting in "literally the same place."

"I saw a shadow on the road, because we drive the same route every time we drive," she said. "I saw a shadow that looked like it, and then something ran into the trees, but it was … all I could really tell was that it was winged and very, very big, but I couldn't give you a definite shape."

The Singular Fortean Society members would make two trips to the site of their encounter.

The investigators didn't notice any radio interference either time. However, upon review of the footage they recorded during their initial visit, they did notice some very unusual audio interference on both cameras used during the investigation. The audio would crackle and pop in an unnatural way within a small radius around the sighting location. As they drove through the intersection where the winged creature had swooped at the two women, the static would begin, only to end a short time after they left and start over when they turned around and passed through again.

The same interference was present upon review of the footage taken on their second trip, although it was much lighter. Just like before, the interference was only recorded in the immediate area of the sighting. They were unable to find any signs of large satellite dishes, ham radio towers, or other equipment in the area that might be responsible for the interference.

OLDER MIDWESTERN SIGHTINGS

Meanwhile, the sightings spread outward not only geographically but also temporally. The longer the investigation into Mothman sightings in the Midwest continued, the more older cases surfaced.

The oldest so far to date came to The Singular Fortean Society in 2020 from Gerald Turrise, who was seventy-nine years old when he submitted his report. Gerald told the Society that he'd seen a winged humanoid in the winter of 1957.

"I realize this story is a little late, but I never knew who to tell my story to until I came across [your article] about the man who saw a birdman at O'Hare Airport," Gerald wrote in an email. "I'm seventy-nine years old now and this happened a long time ago. It was around 1957 when I was with my late brother Gene and his late brother-in law, we were hunting in the Braidwood, Illinois, area for pheasants and rabbits. As we were walking in this large open field, there was a lone large tree standing in the middle of the field. When we were directly under it, a huge man-sized creature sailed over our heads into the woods across the road. We were stunned; we just looked at each other, too dumbfounded to speak."

A follow-up email from the Society asked Gerald for any additional details he might remember about his experience.

"I was a young teenager at the time, and it was early wintertime, [the time of day] was midmorning," he responded. "The winged creature had the body of a large man with legs, but it was covered over the whole body with dark, tan-colored feathers. This happened so long ago I don't recall what the face looked like. I'm glad to tell this story to someone who has knowledge of such creatures before I die."

When asked if he'd had any other paranormal experiences either before or since his sighting, Gerald said that he'd "had an encounter at a US Army station at a Nike Hercules missile site in Northfield, Illinois" in 1963.

The missile site at which Gerald was stationed—which was active from 1955 to 1974—is just south of Chicago, and approximately seventy miles northeast of Braidwood.

"I was on guard duty at the time at our radar station," he elaborated. "It was after midnight when I witnessed a UFO in the sky southeast of my location, maybe a mile away around one thousand feet above the terrain. I watched it for a few minutes before I made a phone call to our command offices in Arlington, Illinois, to report what I was observing. I was contacted a few days later by Dr. J. Allen Hynek. This report is documented in Project Blue Book. Dr. Hynek had told me he was informed to tell me that what I saw was just a private airplane towing an advertising sign. I doubt they would be flying at that time of night. Now, there's no way a plane could ever make the moves I witnessed, and at the end of the encounter, this UFO just shot up into space as fast as a bullet."

Gerald described the UFO as saucer shaped with rotating lights that encircled the craft, illuminating what looked like windows. "[The size was] difficult to know exactly, but I'd say about a hundred feet in diameter," he said. The craft's color, he added, was indiscernible because it was nighttime.

A review of the Project Blue Book case files confirmed the date of Gerald's sighting as May 11, 1963, and the official evaluation as "aircraft."

In further correspondence with The Singular Fortean Society, Gerald explained the impression he got was that Dr. Hynek was being pressured into unfairly discrediting his sighting.

A few days after his sighting, Gerald said someone claiming to be Dr. J. Allen Hynek called him on the phone.

"Well, it was, I think, days later—let's not forget fifty-seven years have passed—and it was late at night," he clarified.

"I was asleep, and the phone woke me. The man [on the other end] introduced himself as Dr. J. Allen Hynek and went on to give me his message that [he said] was relayed to him. He went on to say that he was told to tell me what I witnessed was a small private airplane that was flying over the area, and the plane was towing a sign behind it at around the time of [my sighting].

"He concluded with these words: *This is what I was told to tell you.* I went on to say that there was no airplane that could maneuver the way this UFO was, up and down, side to side, and coming to a complete stop. [There were] lights that completely circled the craft that were rotating, displaying what looked like windows, and the most shocking part of the whole episode was when it departed, the craft looked like it got shot out of a cannon.

"I mean to tell you I was a radar operator at a Nike Hercules missile site, and I've witnessed missile firings, and this UFO took off into space faster than the missiles I've seen being launched. He ended with this: 'I'm sorry, this was all I was to say to you, good night.'

"That was all I can recollect from the past of my account of the incident."

Dr. Hynek was a scientist and UFO investigator who was initially skeptical of the phenomenon when he signed on as a scientific consultant to the United States Air Force's Project Sign in 1948, but his opinion on the quality of evidence in favor of UFOs gradually shifted as he worked on projects like Sign and Blue Book.

He quickly grew frustrated with how flippantly his fellow scientists treated UFOs.

"Ridicule is not part of the scientific method, and people should not be taught that it is," Dr. Hynek wrote in an article for the April 1953 issue of the *Journal of the Optical Society of America* titled "Unusual Aerial Phenomena." "The steady flow of reports, often made in concert by reliable observers, raises questions of scientific obligation and responsibility. Is there ... any residue that is worthy of scientific attention? Or, if there isn't, does not an obligation exist to say so to the public—not in words of open ridicule but seriously, to keep faith with the trust the public places in science and scientists?"

Dr. Hynek went on to eventually found the Center for UFO Studies (CUFOS) in 1973 and even presented a statement on UFOs to the United Nations General Assembly in 1978. He remained a leader in the field of ufology until his death in 1986.

Dr. Hynek resented what he saw as Project Blue Book's mandate to debunk UFOs. While it's impossible to say who exactly told Gerald what on that night in 1963, covering up a legitimately unexplained sighting would have been consistent with the perceived aims of the project. It makes sense to speculate that perhaps someone had ordered the scientist to do so.

CURRENT MIDWESTERN SIGHTINGS

But there's nothing that the government or anyone else can do to suppress anomalous phenomena. For better or worse, the impossible seems to be here to stay. It's not just older cases that continue to filter in as the investigation becomes increasingly well known; new sightings, although reported at a less frantic pace than in 2017 and 2018, continue to be submitted. As of the time of writing, the most recent report comes from Matt Sexton, a truck driver in Illinois.

For Matt, his experiences with Mothman began with two sightings around April or May of 2023.

Both encounters took place in the early morning hours, sometime between 1:30 and 3:30 a.m.

The first time Matt said he saw the creature was on US Highway 30 near the town of Hinckley in Illinois.

"I was driving my truck at the time because I'm a semi driver, so I kind of do later night runs," he said. "And going down that road, I saw a huge branch fall off a tree and kind of, like, hit [the road] a little way ahead of me. So, I looked up and there was this maybe twelve-foot creature. I'm not sure really how to describe it other than it was covered in like a mouse-brown, short-hair fur. It didn't look leathery." The creature, he said, was humanoid.

"It had legs and arms separate from wings, and it came down and went right in front of my truck. It was kind of like when those eagles fall and then they unfurl their wings at the last minute and fly off," Matt said. "That's what it did."

At first the startled truck driver wondered if maybe what he had seen was simply a large bird, but that explanation didn't exactly seem to fit.

"It freaked me out, but I was like, there has to be something I'm missing here," he explained. "And then I just kind of dismissed it."

It became more difficult to dismiss when he saw the same thing only three weeks later, this time on I-39 off US Highway 30, just ten miles or so from his first sighting.

"Again, it went real low in front of my truck," Matt said. "It was the same sort of creature, and I got a better look at it this time."

He described the creature as having a sort of webbing between its legs, almost like a flying squirrel or wingsuit. Its feet had three big claws in front and a talon in the back, and its face had an elongated snout.

"The closest thing I can think of is a gerbil snout or face," Matt said. "It was elongated like that." The creature's wings were membranous and appeared to be separate from its arms, he added. Matt estimated the creature to be at least ten feet tall based on its size relative to nearby reference points. The being's wingspan was similarly huge, Matt added, estimating it to be between twenty and twenty-five feet.

Just as before, the winged humanoid swooped down and then flew back up before disappearing into the darkness.

Again, Matt tried to rationalize his experience, thinking, "Maybe I'm just tired, maybe I'm not seeing right." He wondered if he had simply seen an owl or even just garbage blowing in the wind.

"I kind of went through everything it could have been," he said. "And I just kind of wrote it off because as much as I'm open to paranormal and cryptic things, I'm a fairly skeptical person."

But then he saw it again a year later.

Matt's third encounter took place along McGirr Road about three or four miles west of Kaneville at approximately 2:15 a.m. on April 29, 2024. As he drove his truck onto a small bridge spanning a creek, the creature suddenly appeared.

"I don't know if it came out from under the bridge," Matt said. "It just swooped down real close, almost touching the windshield. And that made me swerve because [the creature] took up so much of [the windshield], and it was so huge. I swerved into the oncoming lane, hit the gravel, and I got my composure back. After that, I was wondering what the hell was going on, just still going. It came down again from the right side to the left over my hood, buzzing over. And then it was thirty, forty yards to the left and up in the sky."

The winged humanoid appeared to Matt as a well-defined black silhouette while it flew. The sky was cloudy, and the lights of nearby Sugar Grove helped illuminate the area just enough to see it silhouetted against the cloud cover.

"I started to speed up and now I hit the gas just trying to get away from whatever it was. And it went over the roof of my truck and then back over the hood again. And it was keeping up with me. I was going about one hundred ten miles an hour, just trying to get as far away as possible."

That's when Matt went for the Taurus nine-millimeter pistol he had with him.

"At this point, I've had too many good looks [at the creature]. I don't know what it is. It's not anything I've seen before, and I'm genuinely fearing for my life. It kept coming down at my vehicle, and when I saw it was keeping up with me, I started firing at it."

Matt was in fight-or-flight mode and thought, "If it's a bird, and I overreacted and killed a bird, then I can deal with that. But if it's something that I need to defend myself from, then, I'll do what I have to.

"Honestly, I'd rather not lose my life or possibly get into a really bad accident over whatever the hell this thing wants to do. I'm not really proud of that necessarily. I'm not like super happy about it, but that's the level of seriousness that it was. I like to think of myself as a somewhat level-headed person, not [someone who would] start firing at any shadow that I see coming out of the woods or whatever. It was genuinely terrifying. It was genuinely something large and completely unseen before that was coming down at my vehicle."

He fired seven or eight rounds at the winged being but doesn't know if he actually hit it.

"It wrapped itself into its wings and started to descend. I followed it down firing, and it was gone after that," Matt said. "I haven't seen it since." The creature had once again disappeared into the darkness.

"I still go by the same places," he continued. "I try to keep an open ear for any weird sounds or anything, and I'm trying to figure out or even process what it was that I saw and what it was that I went through and ultimately had to defend myself from."

The identity of what exactly Matt encountered remains a mystery.

He described the being as seeming to be a biological entity, not exhibiting the glowing red eyes or other potentially paranormal qualities sometimes described in winged humanoid encounters.

However, in what could be seen as one of the synchronicities thought to follow these sightings, Matt mentioned that he had installed a dashcam in the semitruck he had been driving during his first two encounters, only to have the sightings stop after he did so. Furthermore, the dashcam he had installed in his personal vehicle, the truck he had been driving last Monday, had malfunctioned and been removed just two weeks prior to his most recent experience.

"I don't really dismiss any of this just offhandedly," Matt said of any potential paranormal hypotheses. "I'm super skeptical and need to see a lot of evidence. But I absolutely don't just cast any of the paranormal stuff aside. Who knows, right?"

Besides, he noted, "Thinking about a biological entity, I don't think something that big with that kind of wingspan and humanoid build would naturally evolve or would even really be able to fly like it does in the first place."

CHAPTER 6
THE O'HARE CONUNDRUM

Of the reported window areas associated with Mothman in the Midwest, perhaps none are as popular, and problematic, as Chicago's O'Hare International Airport.

Fans of Forteana will already be familiar with the airport. In 2006, a dozen United Airlines employees, corroborated independently by other witnesses, reported a disc-shaped object hovering over O'Hare. The Federal Aviation Administration tried to explain it away as a weather phenomenon, submitting that the pilots, ground crews, and other professionals had been fooled by a circular hole in the clouds. This explanation proved to be unpopular among UFO advocates, and the case remains one of ufology's best-known mysteries to this day.

More recently, it is reputed to be a stomping ground for Mothman.

Over a dozen sightings of flying humanoids and related phenomena have purportedly been reported out of the airport, with considerable variability in their credibility. Many of the more sensational reports, featuring supposed testimony from respected figures like pilots and firefighters, or relating to UFOs and government conspiracies, have come from a single source, Manuel Navarette of website UFO Clearinghouse. Investigators into the phenomenon are dubious of these reports; to date, none of the suspect witnesses' claims of employment or other aspects of their identity have been verified, let alone whether they actually witnessed a monster at America's busiest airport.

The more outlandish sightings deviate substantially from what is usually reported, including details like the appearance of suspicious government employees who turn up to threaten witnesses' livelihoods or UFOs picking up Mothman in a beam of light. This deviation, combined with the inability to prove that the people making these claims even exist, let alone are who they say they are, has left plenty of room for doubt as to the veracity of the reports. Doubt that would likely be grounds for outright dismissal of the testimony, if it wasn't for the handful of reported sightings that exist with a more generous portion of credibility.

DANIEL'S SIGHTING

In late 2019, The Singular Fortean Society was contacted by a man who said that he'd seen a red-eyed winged humanoid while driving near O'Hare on the night of December 6, 2019.

The witness, who agreed to the use of his first name, Daniel, said that he works in cargo for an airline in the airport and had left work early that night to go meet friends.

"I was the last one to get out of work," he said. "I normally work until 11:00 p.m., but we got out a little earlier than expected. Around 10:15 or 10:20 p.m. is when I left. I [was driving] down West Higgins, and I was headed to the bar to meet up with one of my friends from work. [To my left, in a large patch of grass] is where I saw it.

"It was dark, and there's a building with a certain amount of light [near the sighting area]. I was able to take a glimpse at it while I was driving. I saw headlights coming toward me as well. I'm always looking at that side because there's deer there. I'm always eager to look there to see them. I can tell the difference between a deer and what I saw. What I saw was not a deer."

According to Daniel, the creature he saw matched exactly what was described by a witness who said that, just over a week before Daniel's sighting, he had seen a seven-foot-tall "person with wings" near the loading dock for Nippon Cargo Airlines. Except the thing he saw had red eyes. He said he was able to estimate the creature's height by comparing it to a nearby fence.

But the similarities didn't end with a physical description; present, too, was the feeling of evil described by the previous witness.

That first witness was a man who said he was standing outside of a cargo dock at O'Hare when he spotted a seven-foot-tall "person with wings" just outside of a fence by the parking lot. The sighting reportedly took place at approximately 6:30 p.m. on November 26, 2019.

The sighting was investigated by Navarette, and according to the report he received from the man:

"I was at the airport picking up a load at Nippon [Cargo Airlines]. I was already backed into a dock and was standing away from the truck smoking a cigarette while they loaded my truck. I was looking toward the runways, in the direction of the tunnel, and that is when I noticed something that looked like a large bird standing just outside of the fence by the parking lot. It was not hard to miss because two streetlamps were nearby. It looked like a person with wings that were stretched out and flapping. It was walking away from the fence toward the open field and then began to flap its wings and disappeared."

According to Navarette, he was able to speak with the witness over the phone.

"I spoke with the witness via phone and was able to get a little more information regarding this sighting," the investigator said. "The witness primarily speaks Spanish but was able to report this sighting with the help of his daughter and her boyfriend."

"[He] was standing away from his truck as it was being loaded, smoking a cigarette, when he said he caught movement out of the corner of his eye and saw the being standing near the parking lot [that] was illuminated by two streetlamps. The witness stated that the creature was about seven feet tall using the fence as a point of reference," Navarette explained. "When I asked him how he was able to be so certain as to the height of this being, the driver stated he has been to this location multiple times and he estimates the fence to be about eight feet high. Using the fence, he was certain that the being was at least seven feet tall. When I asked him how large the wings were, he said at least six feet across and black."

The witness used language also seen in earlier sighting reports from the largely Hispanic neighborhood of Chicago's Little Village.

"When I asked him to describe the being, he said it looked like a *demonio* (demon) or a Duende (goblin) and was solid black. The witness said he saw nothing that looked like eyes and he assumed the creature might have had his back turned to him. He stated that it walked with a gait like a bird and that it was flapping its wings as it walked toward the large field that is by the runways, and disappeared into the night," Navarette said. "The witness did state that when it disappeared, he quickly did the sign of the cross and asked the Virgin Mary for protection. He put out his cigarette and quickly walked back to his truck. When I asked him why he did that, he stated that he felt a presence that was evil and was convinced that he had seen a demon. When asked to elaborate on this statement the witness refused to talk about it anymore for fear of it coming back."

It's worth noting that this repeated language is something critics use as evidence of possible hoaxing in the reports. Navarette is himself Mexican American, and there are those who find it suspicious that the reports he provides so often include elements of that culture. However, as others have noted, he isn't alone in receiving reports from members of the Hispanic community, and if he has received more than other investigators, it could be due to his connection to the community.

When asked if there were any other witnesses, the truck driver reportedly told Navarette that "there were others at the same facility, but many were either inside the facility itself or in their trucks."

And, apparently, this wasn't the man's only sighting of such a creature in his lifetime.

"When asked if he had seen something similar before, the witness stated that he had when he was a teenager back home in Mexico," Navarette said. "The witness stated that he saw a solid black winged creature that was circling an open field that he and other children were playing soccer in. He stated it circled the field and made a loud screeching noise before flying off into the surrounding forest. When I asked him if he remembered the date of the sighting, he stated that he did not remember the exact date, but a week later there was a large earthquake in Mexico City. For the record, the magnitude 8.0 earthquake that hit Mexico City was on September 19, 1985."

Navarette ultimately concluded, "The witness seemed sincere, albeit scared that he had seen something demonic and evil. It is my opinion that the witness is telling the truth."

As for Daniel, he was ready to dismiss his sighting until he found an article online describing the other man's experience.

"That was really weird because, honestly, I thought I was going crazy at first. I said to myself, 'No, that can't be it. I don't think I saw that,'" he said. "After that, once I saw the article [the day after my sighting], I thought, 'Okay, I must not be going crazy, because Nippon is not far from where [my sighting took place].' That was the craziest part. I kind of freaked out. I saw that something wasn't right. I saw it the same way that the other gentleman saw it. It felt really evil. It had those red eyes. How could any creature have those?"

"I thought I was going crazy, man. I thought I was going crazy, until I read the article. I must not be the only one who's seeing this stuff," Daniel added. "Maybe it was just the fear I felt, but honestly, it didn't feel good at all. It was one of the creepiest things I've ever seen in my life."

Lon Strickler also interviewed Daniel, and afterward told The Singular Fortean Society that he found the man's testimony credible. Navarette was the only investigator to interview the witness who inspired Daniel to report his sighting.

This was a recurring issue with the sightings coming out of O'Hare, and in fact, only one other sighting report involved contact with multiple investigators. Unfortunately, that sighting would prove more problematic than the one reported by Daniel.

TOLLBOOTH SIGHTING

In early October 2019, MUFON received a report from a man sitting in his car outside of O'Hare at around 9:45 p.m. on the night of October 5 when he spotted a "tall creature" with "bright red eyes" and "large wings."

He claimed he was a rideshare driver and that, while parked on the side of the road near Toll Booth Plaza 31 awaiting his next fare, he "saw something walk out of the trees."

"It was a tall creature about six to seven feet tall and kind of hunched over," the man wrote in his report. "It walked with a sort of gait like almost a waddle. It was completely black; when it swiveled its head, I saw two bright red eyes glowing."

As he watched, stunned, "This thing opened a pair of large wings, as wide as it was tall, and began to flap them. It took off and headed to the south toward that large indoor driving range and the baseball park. It was gone in a few seconds."

The Singular Fortean Society made contact with Illinois MUFON State Director Sam Maranto shortly after the report was filed.

Maranto said that a MUFON investigator had exchanged several emails with the witness, but that the witness "is refusing to talk on the phone or meet them." He added that the investigator following up on the case "knows the location very well and will be interviewing folks around the site."

The demeanor of the witness seemed odd, Maranto said, since "most folks are anxious to talk, give more information, or help in any way."

The MUFON state director offered to keep The Singular Fortean Society informed of any new developments in their investigation, although there would prove to be no new developments on their end.

Shortly thereafter, Lon Strickler told The Singular Fortean Society that the witness had contacted him by email after deciding not to speak directly with MUFON. According to Strickler, the witness said he was contacted by a MUFON representative who "told him that he [wouldn't] believe him unless he came forward . . . with his name and a photo of himself."

Maranto denied that the investigator assigned to the case made the alleged requests and stated that most of the questions sent to the witness were taken directly from the standard MUFON interview questionnaire—the only deviation being a request from the investigator asking the witness to review photographs of the sighting area to provide more details.

Strickler would go on to speak directly with the witness, who told the investigator that he decided not to communicate further with the MUFON field investigator assigned to his case after the investigator displayed a "pushy and aggressive posture."

The man also added details about his sighting experience, telling Strickler that with his window rolled down he was able to detect the strong scent of ammonia after the creature appeared.

"It was very pungent but not overwhelming," he said. "When it took off, the smell was gone within a minute [or] two."

And though the scent of ammonia may have faded, to many investigators, something still smells off about the claimed sightings at O'Hare.

CHAPTER 7
THE HIGH STRANGENESS MAN

Joshua Cutchin is one of the last true Forteans. He's written books on scents associated with various paranormal phenomena, the similarities between faerie changelings and alien abductions, and high strangeness surrounding Bigfoot sightings, among others. When Richard and Tobias were able to sit down and talk with Joshua about such anomalies, they had some idea how the conversation might go, but the two investigators could never have predicted where it would eventually lead. Naturally, it began with Charles Fort.

Joshua's introduction to Fort came in part through his ubiquity among certain researchers, such as Bob Rickard,

Jerome Clark, and, of course, John Keel. Researchers who, although adamant in their insistence of Fort's relevance, have still failed to make him a household name even within the paranormal field. That's not their fault since, as Joshua pointed out, it takes a particular type of person to embrace the Fortean perspective.

"Once you break the seal on Charles Fort, he bleeds into your life; if you're into these things and you're of a certain disposition. It strikes me that there are exceedingly few Forteans nowadays and exceedingly few people who aspire to be Fortean nowadays," Joshua said. "With Fort, there is a certain resistance to categorization and conclusion that I don't see across a lot of modern paranormal thinkers. There's this drive toward certainty to be proven right about one's pet theory, and that's the antithesis of what Fort was dealing with."

Fort's writing is decidedly inconvenient to those with preconceived notions of how the universe works, which can make it unpopular with a certain crowd, despite his many contributions to anomalous research.

"[Fort's work is] just this presentation of things that happen, and here they are, and I can maybe cobble together a worldview that accommodates them, but at the end of the day, you're just left throwing up your hands and saying, these things happen, and people are reporting them, and here are those damn facts. But what I really think is interesting about

Fort is how he was on the forefront, as you alluded to, of so many things that we're still dealing with today, like the coining of the word *teleportation*, or the manner in which reality seems more malleable," Joshua explained.

He spoke further about Fort's influence on the field, noting the early twentieth-century writer's prediction that the paranormal would inevitably be included into mainstream society following the stagnation of science and religion only to be similarly replaced by something new once this new paradigm itself inexorably grew inflexible. This process, Joshua said, has been playing out over the last several years, following a decades-long cycle that seems to endlessly repeat itself.

"Fort was dealing with some aspects of things that sounded a lot like simulation theory, talking about being emanations from X, that idea that we were emanations from another mind in another place, characters in a novel. And I think these are all really valid things. That is something I have tended to notice about people who have ideas that are truly timeless and are truly durable, is they can not only be read in the context of the times, but they also seem to have shaped the times as well. And that's part of Fort's legacy," he said, before steering the conversation toward friendly debate. "But speaking of that legacy, I think you're right to draw Keel into the discussion. I know, Tobias, that you've expressed less-than-warm feelings about John Keel in the past."

This wasn't a surprise to Tobias, who responded amicably, "Keel is one of the most influential writers in my life, and like most things, unfortunately, when you gain new perspective as an adult, a lot of the shine comes off him. I still think he's one of the most entertaining writers, certainly in this space, but really just in general, that I've ever read. But when you scratch the surface of a lot of what he's saying, you find that it either is all surface or it's smoke and mirrors. It's an illusion. None of it exists."

Keel's veracity is of less importance to Joshua, who said, "Well, first of all, let's take a look at this though, in the context of Fort, because Keel being in that lineage is a natural progression with the way that Fort himself was drawing on older things and things of the time. There's a certain pulpiness that you can see both across Fort being an influence on pulps, and then John Keel being sort of the pulp paranormalist. But I would also say that I think that John Keel should probably be recognized as the originator of gonzo journalism, really."

He continued, adding, "And this is something that Nick Redfern said, and Greg Bishop has said, but Keel reported on these things with that complete lack of objectivity, which is odd because he was a science editor for Funk & Wagnalls, and you don't get that gig without knowing how to be objective. But he still, for whatever reason, found it important to include his own voice in a very personal way in all his writing. I do think it's interesting for me as I cycle through new ideas that I'm really proud of, I'll oftentimes pick up a book by Keel and reread it because I read these things a half dozen times and am like, oh, no, that's not my original idea. I just cut that from Keel, who in turn probably picked and chose from Fort as well. So, I think that's important to acknowledge too."

And in some way, it could be Keel's very lack of reliability as a narrator that makes his work so readable.

"I do think there's some additional utility in the way that Keel makes some of those ideas more approachable, right? Because you're not going to find somebody on the bus reading *The Book of the Damned*. It's just not going to happen," Joshua said. "And part of that's a function of the time in which it was written, and I love Fort's writing. I really do. But there is a sort of elegance there. There's an elegance to Keel, I think made a little bit folksier and a little bit more palatable to a lot of people. So, I think that's another aspect of his legacy that should probably be preserved in a better light. Sure. But you're right, there is a lot of smoke and mirrors there."

Tobias and Richard both readily agreed with Joshua's assessment of Fort's appeal to the layperson.

"Exhibit A for the defense," Richard quipped. "I totally agree with you. I've read those books at Tobias's suggestion, and I ask myself, What did I just read? I got to go through that paragraph again and deconstruct it."

For Joshua, Keel's utility comes from his ideas, rather than any "facts" he may have claimed to share.

"I don't go to John Keel's work as a chronicle of exactly what happened. I find myself returning to John Keel's work because of the ideas and the concepts that he was putting forth, which I still think do have some utility," he said.

"And honestly, with the way that these phenomena seem to behave, whether or not they were accurate, they've sort of become accurate. You know what I mean? It's almost like if we incorporate sort of a Fortean perspective about how perception does influence the way things really are, then you could say some ideas that Fort had have come to be true.

"Not that he was accurately perceiving them at the time, but they have so much influence on paranormal culture that we have adopted those. On the more skeptical side, you say, yeah, we adopted them and they're still not true. But I don't know. You look at things like egregores and tulpas and thought forms and just, again, the way that reality does seem to be fluid, and you wonder if he didn't seed some ideas into the paranormal consciousness that have found evidence for themselves as opposed to being accurate assessments of the evidence at the time for him, if that makes any sense."

Tobias laughed, "So, I'm going to draw a parallel between you and John Keel right now because you're doing a classic John Keel maneuver, and that is taking something totally weird and completely unproven and talking about it as though we all know that it's a real thing. When you say stuff like, 'Well, think about tulpas and egregores.' I'm like, 'I do all the time,' and yet I've never seen one."

Joshua smiled, responding, "I totally get what you're saying. But again, it starts to become sort of a 'turtles all the way down' thing, because I can say, 'Okay, I'll see your refutation of egregores and tulpas and thought forms and raise you the New Age movement and the New Thought movement and powers of manifestation and occultism.' It's like there is an almost universal belief in things like manifestation and the idea that perception does influence reality. That's an old religious tradition.

"So again, without casting judgment on whether or not any of those things are right or wrong, the tulpa thing is a problem. It's a huge problem in the paranormal. It's all appropriation, and it's based off Alexandra David-Néel, and that's about it.

"I spoke with someone who said that they had an audience with a high-up lama, and they were really excited because they're like, 'I'm going to ask them about tulpas.' And they asked, 'Can you tell me more about tulpas?' And it went through the translator and the lama goes, 'What?'

"I guess what I'm getting at is there is sort of a universal belief of things like belief shaping reality, and that's more of a philosophical question, which is what I was trying to propose as the spirit of Keel's influence on the paranormal, but I gave you two options. I gave you the skeptical version, which is confirmation bias. Right? Keel says these things, and then we start to use confirmation bias to make them appear true. That's on one end of the spectrum. The other end of the spectrum is that he said these things were true, and the phenomenon has slowly adapted to it."

Tobias nodded, saying, "Sure. Which I think is a wonderful idea, and I think it's cool to be open-minded. But yeah, I mean, it's not like we have evidence.

"Let's assume for the sake of argument that the effect Keel's influence is having is weighed heavily toward confirmation bias. What is the overall effect of that? Because from my perspective, that's a huge issue. Essentially what that leads to is people being led astray. You move farther away from the truth, the more you succumb to things like confirmation bias."

Joshua responded, "And that makes perfect sense given the regard with which you hold for that. You would feel that way because you are embodying that sort of Fortean agnosticism. But you do reach a certain point if you follow that trajectory where you start to call all these things into question, and I know that you do [question things]."

Still, he continued, "Ghosts might not be dead people, but we just carry along our merry way that they are. Same thing with the UFO question. They probably aren't aliens, but we carry along like they are, because that's the only way we have to deal with these things.

"It really becomes a question of where every investigator or person or witness chooses to draw the line. I think now, as far as that confirmation bias, it is important to realize that Keel wasn't living in a vacuum. He did invoke the idea of tulpas. His idea of the super spectrum looks a lot like Jung's collective unconscious.

"So, he was sort of engaging in this cafeteria-style speculation where he was taking bits and bobs from here and there as he best saw fit. In that case, people sort of adopting those stances today is entirely consistent with what Keel was doing for better or for ill."

"Fair enough," Tobias said. "How seriously do you think Keel took his own hypotheses, and how seriously do you think he expected other people to take them? I go back and forth on this all the time. I still don't know if he really believed in Ultraterrestrials."

Joshua thought for a moment before saying, "I don't know. Even if you're a fan of John Keel, you do get the impression that he's throwing a lot of things at the wall to see if they stick. And I feel like I'm in the lion's den here,

Tobias, because I see that as a virtue, but I know you probably don't. But yeah, I think that in this domain sometimes the best approach is to just throw a bunch of different ideas against the wall and, quite frankly, end up contradicting yourself. That might be the only way to build a true legacy in this field that is timeless, to just keep on proposing stuff and hope that one of your ideas is right, or ..."

"Or at least it's popular," Tobias suggested.

"Or at least it's popular," Joshua agreed.

"I will say that there are some things that Keel seemed to be especially prescient about. He compared UFO contact to what he was calling death dreams, and the reason that he was calling them death dreams is because the term *near-death experience* hadn't been invented by Raymond Moody yet.

"But he was already drawing those connections before they came out with Kenneth Ring's *Omega Project* in 1992, which was viewed as sort of revelatory by the UFO community. It's just another one of those things that are sort of buried in Keel, and it might've been one of those ideas that he was just throwing up against a wall that just happened to stick.

"So as far as how much he bought his own BS in the Robert Anton Wilson sense of buying your own BS, I'm not sure. I don't know enough about his personal life to really evaluate. I do think there's an important clue in the fact that he was a child fan of stage magic. That feels really important. Is this all one big sleight of hand by John Keel? That's a question you have to sort of sit with."

Keel was a self-avowed writer of propaganda while serving in the US Army during the Korean War, but he was also an iconoclast with a deep distrust of authority. That distrust of authority, Joshua explained, can sometimes be a destructive influence within the field of Forteana.

"There's this knee-jerk impulse that a lot of Forteans have to authority and structures that are imposing their will on your version of reality. And don't get me wrong, I very much sympathize with that, but at some point, you have to realize that, yes, there are people lying to you, but not everybody is lying to you. But Fort taken to the extreme... you do wind up in flat earth territory," he said. "You have these excerpts from Keel where he's talking about puppet masters and being manipulated and all these other things, which again, I can't cast judgment on whether or not that's actually what's going on, but it's a pretty dark way to move through reality."

However, Joshua continued, "It occurs to me that sometimes it's best to engage with some of these ideas, and then when you start to feel yourself slipping off the edge past the point of no return too abruptly, pull yourself back to consensus reality. Maybe that's what he was doing by putting a lot of different ideas against the wall; he was going to some of these darker places and saying, 'Oh, this is a little bit too weird for me. Maybe let's try something else.'"

Richard quickly jumped in here to add, "Keel speaks with this blithe confidence that everything he says is true, just trust me. Yet I see no citations. So, *The Mothman Prophecies* is literally 296 pages of 'take my word for this.'"

Naturally, Tobias agreed. "If Keel is kidding, he doesn't make it obvious at all."

All three men agreed that, in their own writing, they tend not to speak with such certainty, especially since they have a responsibility not to mislead anyone. None of them know precisely what is behind the phenomena that people report and pretending otherwise is dishonest. In this sense, Keel departs from Fort, since while Keel often proclaimed certain things to be true or false, Fort rarely did, instead being more inclined to wonder: What if?

Moving on, the conversation turned to the relative importance of objective reality. Joshua hasn't had many of what he would call paranormal experiences, and as such, takes a different approach to them than does Tobias, who has had many. For Joshua, the truth is not necessarily important, and he is content for these phenomena to exist in a state of indeterminacy, but for Tobias, the truth behind what has happened to him and countless others is of the utmost importance. While Tobias has accepted that nobody knows exactly what is behind paranormal phenomena, that doesn't mean we shouldn't try to understand them, even (or especially) when that means identifying falsehoods or other prosaic explanations.

It is here that Richard mentioned the seeming lack of physicality to the Mothman phenomenon.

"That's another aspect of Keel that for all its darkness I think is worth emphasizing—his notion of a superposition intelligence pulling the strings is sometimes the best way to look at these phenomena," Joshua responded. "It does have a sort of 'God of the gaps' flavor to it, but if these things that Keel is reporting as true should be taken as true in the literal sense, there are no bodies because these things dip in and out of our dimensions. There are bad photographs because this thing can pull the strings of the way film develops."

To which Richard replied, "Coming to this as a complete newbie, if this were the plot of a novel, then Keel's Ultraterrestrial hypothesis seems like a deus ex machina to me. It is the ultimate 'get out of jail free' card for any criticism you level at it. No pictures. Well, we got something for that. No physical evidence. Well, they look the same but different. That doesn't mean it isn't necessarily true or doesn't hold water. It just means that sometimes reality is more poorly scripted than fiction, if that makes any sense."

This comparison to fiction seemingly reminded Richard of the infamous Silver Bridge collapse so often tied to Mothman. "This is what led me to the conclusion that this maligned link with the Silver Bridge collapse screams of someone who has no third-act finale for his narrative."

Tobias agreed. To him, the addition of the Silver Bridge in the Mothman narrative has long seemed a contrivance made more to sensationalize the phenomenon and provide a convenient ending for Keel's book.

Joshua tied this to their earlier talk of confirmation bias, saying, "It's a very good example of the paranormal community accepting these things as fact, and then in some sense, turning them into fact, right? Because even if there was no real association at the time, in the zeitgeist of Point Pleasant between Mothman and the collapse of the Silver Bridge, that's certainly become the Mothman's modus operandi since then—that it appears before tragedies."

In rebuttal, Tobias pointed out that, in fact, since the Silver Bridge, there have been no credible Mothman reports from before any tragedies or disasters. They only ever come after, which likely indicates that this is a narrative invention probably started by hoaxers but carried by believers. People all too willing to perpetuate a falsehood—knowingly or not—if it fits into their idea of how the phenomenon is supposed to behave.

"Now, I will say, I think that Mothman and being a harbinger of disaster actually argues against this phenomenon mimicking any narratives we create, subconscious or otherwise," Tobias said. "Because the reality of these sightings when you work with witnesses, and I've spoken to dozens, is that none of these sightings have anything to do with any disasters. It just never happens. It never comes up. So, if us believing in that is supposed to influence the phenomenon, it isn't."

This became another point of general agreement among the three researchers. Believe it or not, Tobias and Joshua are friends with a deep mutual respect and do generally agree more often than they don't. But even in their disagreements they still respect each other. And, of course, everybody likes Richard.

Ultimately, what they could all agree on is that a healthy dose of open-minded skepticism is probably best for approaching these subjects, and in at least the case of Mothman, it's okay to not care too much about the truth if all you're interested in is a good story. But never lose sight of the fact that these are real events happening to real people, sometimes with tragic consequences.

When it comes to the events in Point Pleasant, Joshua concluded, "The tragedy [of the Silver Bridge] should be the main focus. And if you want to bring the paranormal into the discussion, I think it should be framed within that context, not as the paranormal necessarily causing it, but as the tragedy being so traumatic that it influenced events in the community. I think that what *The Mothman Prophecies* and Keel's legacy writ large represent is, however, entirely consistent with what we see across all aspects of the paranormal, which is mythmaking and the cyclicality between storytelling and true events. And this is where I am right now. We've been conditioned to think that that flows one way and I'm not sure that it always does."

Well said, Joshua. Another thing everyone can agree on.

CHAPTER 8
ENCOUNTER AT DUFIELD POND

In their capacity as paranormal investigators, both Tobias and Richard tended to be skeptical of those who came forward bearing claims of strange encounters if they were seeking to publicize their story as much as they possibly could. That wasn't to say that some of those instances weren't genuine brushes with the unknown; however, it did cast the motives of the individuals concerned into some doubt.

One was forced to wonder whether they were trying to cash in on the situation in some way. Some people wanted to be famous (or infamous). Others wanted to see themselves on TV or to make money from it. Those accusations have been leveled at experiencers ever since researchers first started recording and investigating such claims. Sometimes they turned out to be true. Sometimes, false. A lot of the time, the whole truth never came out in its entirety.

In 2019, a case crossed Tobias's desk that made him sit up and pay close attention. The first thing that struck him was that the witness actively shunned the limelight; so much so, in fact, that he hadn't wanted to talk about it at all. Fortunately, his wife contacted veteran researcher Lon Strickler to report the bizarre occurrence.

She told Strickler that the incident happened relatively early on the evening of February 22, at around 8:00 p.m. By then, it was dark. The man was driving back from having visited the pharmacy. Suddenly, his car headlights illuminated something exceptionally strange.

It was a massive humanoid creature, somewhere between eight and nine feet tall. The being was crossing the road and quickly disappeared from his sight.

When he got home, the presumably shaken man told his wife what he had encountered on the road, narrowly avoiding a collision as it passed mere feet in front of his vehicle. She rather sensibly asked her husband to sketch it. What emerged on the sheet of drawing paper was the outline of a powerful humanoid figure, striding forcefully from right to left. Tucked behind it was a set of large wings.

Tellingly, there was no mention of the glowing red eyes, which are a staple of Mothman sightings. Still, the other physical similarities were noteworthy. Unwilling to simply let it slide, the witness's wife went out to the location of the sighting the following morning to check things out for herself. February in Illinois usually means snow and ice, and she thought there might be a chance that the entity could have left tracks. What she found out there was a plethora of boot and shoe prints, interspersed with numerous animal tracks made by the critters that live and hunt in the area.

Nestled among them was a print that, to her eye, did not seem like the others. It was clearly human-shaped but elongated. There was no obvious tread pattern made by the sole of a manufactured piece of footwear. Toe imprints were visible. A skeptic would point out that perhaps it was exactly that, with erosion from the elements wiping out the more artificial features. However, the print had only been exposed to the Illinois weather for a single night, and other tracks in the vicinity showed somewhat more tread. Still, this was far from conclusive evidence.

When Strickler spoke to the witness by phone, the man clarified that the wings had been membranous in appearance, which is another relatively common characteristic in Mothman sightings. Strickler also emphasized that this had been no mere fleeting glance. The driver had caught the entity full-on in his high beams, so this could not be explained as a simple trick of the light caused by shadows and motion.

Strickler also noted that the eyewitness believed the being, whatever it was, to have been some sort of harbinger or ill omen. One possibility for that would be the link that exists in the public consciousness ever since Mothman arrived on the scene and became inextricably—and in the opinions of the authors of this book, unfairly—connected with disaster because of the Silver Bridge collapse.

On the other hand, was this a case of the witness somehow subconsciously picking up on a sense of imminent danger or impending doom? A primal biological or possibly even psychic defense mechanism against something which might be capable of causing great harm?

In search of answers, Tobias set out to visit the site of the encounter itself. The location was named Dufield Pond and fell under the auspices of the McHenry County Conservation District. After parking his car and getting out to survey the lie of the land, he reconnoitered the area and set about putting himself in the position of the eyewitness. The pond was open from sunrise to sunset and was long since closed when the incident took place.

Walking along the gravel track that led from the parking lot to the road outside, Tobias observed that the entrance to Dufield Pond sits just beyond a curve, forcing motorists to slow down slightly as they negotiate it. This reduction in speed would have given the driver an extra split second or two to watch the entity as it crossed the road.

Located directly across from the road were privately owned, single-story residences. The winged humanoid would, presumably, have disappeared into one of their front yards. Alternatively, it could simply have flown off into the night sky. The driver didn't hang around to find out.

Making his way back to the pond, Tobias stood at the edge and surveyed the pool. According to a data chart, the pond was nineteen feet deep in the center. Could the entity have emerged from its depths? Possibly, although it hadn't appeared to be wet and dripping according to the eyewitness. He also wondered why, if the footprint in the photograph sent to him by the driver's wife did indeed belong to it, the creature had chosen to walk the length of the track and cross the road on foot, rather than simply fly or glide.

To shed light on the remarkable sighting, Tobias and Richard came to Dufield Pond together.

An information board listed all the wildlife that called this area home. The pair perused it with interest. Visitors were advised on what to do if they should encounter a coyote—wave their arms, yell, throw stones, and generally act aggressively to frighten it off—and under no circumstances should they turn their back and flee.

Of more immediate concerns were the mosquitoes, which zipped around both men in an ever-present insectile cloud. Every once in a while, a particularly bold one would land on one of their bare arms, forcing them to swat it away.

There were also ticks to be reckoned with, and at the end of the investigation, they would make a point of checking themselves over carefully before leaving. Neither Tobias nor Richard wanted to contract Lyme disease.

As they carefully skirted the water's edge, Tobias emphasized to Richard once again how impressed he was with the credibility of the eyewitness. The driver had never asked for money in exchange for his story. He had actively shunned any and all publicity. There was no motive for him not to tell the truth, and no reason to believe that the driver was either making it all up for kicks or simply delusional.

"It's one thing to read about strange cases such as this," Tobias went on, "but it's another thing entirely to sit face-to-face with another adult and see the emotions they convey when they relate their experiences. As an investigator, you watch the fear that still lives inside them playing out on their face. That's difficult to fake—though admittedly not impossible—and it's not the kind of thing somebody just does for no reason. This witness had a lot to lose by coming forward, and nothing to gain. That's why he was initially very reluctant about ever coming forward."

He made a good point. Tobias interviewed the witness in the man's own home, with his spouse present, and could tell that he was profoundly disturbed by the encounter.

"This upset his entire worldview," added Tobias, "and that's pretty common among witnesses who have had authentic experiences like this."

Some documented brushes with the Mothman involve a sense of intense, almost palpable fear radiating from the creature. Others, however, are more like this case. The fear, the feeling of being deeply disturbed by the seismic shift in their belief system, only came afterward—but lingered for a considerable time, and in some instances, never went away at all. This is by no means a universal phenomenon, but it seems to occur in the majority of the cases, in Tobias's estimation.

"This witness was driving, and his sighting happened so quickly, that I don't think he even had time for fear to register in the moment," Tobias concluded. "The fear set in afterward, when he got home and had time to ponder it. Despite the fear that he might be ridiculed, he took the risk and came forward anyway."

That made both researchers wonder just how many similar cases go unreported, simply because the experiencers don't want to deal with the laughter and finger-pointing that inevitably takes place when such an episode is made public.

They wandered farther along the entrance track. Tobias stopped about fifty feet back from the road, pointing out the spot at which the anomalous footprint in the ice was found.

Thinking skeptically, they considered the likelihood of this very human-looking footprint having been exactly that... a human footprint. Except that didn't quite add up. For one thing, there was only a single print found. The remainder surrounding it were all either animal tracks or had been made by the soles of shoes. The anomaly was found almost but not quite in the middle of the track, which prompted the question: Did an anonymous prankster really step out into the middle of a frozen track, slip off their sock and shoe, then push their bare foot down into the ice with sufficient force to leave a firm imprint? They would then presumably have had to stand on one leg while putting the sock and shoe back on, before making an exit.

Even if the investigators posited that this was possible—as indeed it was—they were still left with the burning question of why in the world anybody would ever *want* to do that? It's not exactly top-level pranking, and this explanation does nothing whatsoever to explain the sighting of the enormous winged humanoid the driver also saw.

Tobias and Richard just couldn't understand why, or how, there would only be a single footprint. The length of the footprint was roughly twice that of Tobias's own, and about 30 percent wider. For comparison, Tobias wears a size 12 shoe.

Standing in the same location, they discussed possible explanations, discarding one after another those which don't seem to fit. Fraud was discounted immediately, for the reasons previously stated; this man had everything to lose in terms of reputation and absolutely nothing to gain by reporting his experience to investigators.

Could his eyes have been playing tricks? The bend in the road, coupled with a 45-mile-per-hour speed limit, means that drivers pass by the Dufield Pond entrance fairly quickly—but not so quickly that someone (or something) walking across the road in front of them would fail to register. Admittedly, the shadows were long at that time of the day, and had this been only a corner-of-the-eye sighting, we would already have dismissed it as a trick of the light or an overly active imagination ... but that isn't satisfactory. The driver reported seeing the entity walk directly in front of his vehicle and passing straight through the beams of his headlights, at which point it was fully illuminated.

If this was indeed a genuine sighting of either Mothman or a similarly winged humanoid, it prompted the question of what exactly the creature was doing at Dufield Pond. Tobias pointed out that in the vast majority of such cases, there is no evidence of predation; in other words, no leftover animal carcasses or remains that would indicate a flesh-and-blood entity had been feeding on nearby wildlife. Nor is there scat, which would represent the opposite pole of the digestive process.

What business would the entity have in the vicinity of a deserted pond, if not to eat or rest? The evidence suggests that Mothman enters our reality from somewhere outside it—*elsewhere,* for lack of a better term—and then returns there in fairly short order.

"It's highly improbable that this is an actual biological creature living in this area," Tobias said thoughtfully. "Indeed, the opposite seems to be true. It's almost as if what we're investigating here is an idea . . . the idea of the archetypal bog monster, somehow brought to life and given physical form."

If that really was the case, then something exceptionally strange is at work in Woodstock. The kind of something that conjures up a winged creature, apparently out of thin air, after which that creature strides confidently across a busy road during the hours of darkness rather than spread those wings and use them to fly. Never seen in the vicinity before—or at least, never reported—the entity simply disappeared, and at the time of writing, has not appeared again.

However, if the Dufield Pond case was an isolated incident, neither Tobias nor Richard would have journeyed all the way out there to investigate it in person. This part of Illinois is teeming with winged humanoid sightings, and the quest to find out more about the elusive Mothman is about to take them to the scene of another even more bizarre encounter.

CHAPTER 9
NIGHT TERRORS

Elsewhere in this book, we have covered Mothman sightings that have taken place in swamplands, nature reserves, and out in the woods. Most of these are sparsely populated, sometimes desolate areas in which there are few people around to serve as eyewitnesses.

Contrast that with the next encounter to be examined, which took place in a residential district in Rockford, a suburb of Chicago … an ordinary street in an ordinary neighborhood that probably looks no different to the one in which most of us live.

Tobias and Richard met Jonathan, the experiencer, at his home. After an introduction was made—Tobias and Jonathan had both met before—they were beckoned inside Jonathan's home. They made themselves comfortable on the couch in the living room, and while Richard set out a pair of digital voice recorders to preserve the interview for posterity, Tobias prepared to start jotting down notes.

The encounter happened several years prior, when Jonathan was in high school. Jonathan's mother worked overnights as a nurse. She typically left for work at around 10:30 each night. Just for safety purposes—Rockford has crime, just like any other city—it became traditional for either Jonathan or their father to walk her to the car, which was parked on the street.

On that particular night, it was Jonathan who saw her off to work. As the car pulled away from the curb and turned the corner, disappearing from sight, Jonathan turned and started back toward the house. That was when they heard the sound. Today, they remember it as sounding akin to "a velociraptor screeching," conjuring mental images of the Jurassic Park movies. The ululating screech was so loud, it ought to have brought out every resident of the neighborhood to investigate, yet not a curtain twitched, nor did any of the front doors open.

Jonathan looked toward what seemed to be the source: a tree located on the edge of the front yard. There was something odd about the tree. Between the boughs, it was black. Not the ordinary shadowy darkness of nighttime. This was a peculiar type of inky blackness, one which suggested that something might be lurking within.

As abruptly as it had begun, the screeching stopped.

That's when Jonathan saw the eyes. Huge, glowing red, they stared back at Jonathan from the depths of the darkness. Jonathan estimated each one as being about the same size as a dinner plate.

No sooner had Jonathan made eye contact than the screeching returned with a vengeance. Adrenaline surged through Jonathan's system, part of the natural fight-or-flight response, which accompanies a close encounter with something unknown and potentially terrifying.

"Something about this entity felt malicious," Jonathan recalled. "I don't know if it meant me harm, but ... there was a form of communication taking place that felt negative to me. I don't know if it revealed itself to intimidate or scare me, but causing fear is what that screech seemed intended to do. It was definitely aggressive."

So frightened was Jonathan that he bolted for the front door, rushed into the house, and slammed it behind him.

Considering the initial encounter with the benefit of hindsight raises an interesting and potentially disturbing possibility: Was the entity perching in the tree when Jonathan first left the house accompanying their mother to the car? If so, then the implication was that it may well have waited silently for her to drive away—perhaps waiting for Jonathan to be alone and relatively isolated.

"Why did it want my attention?" Jonathan asked rhetorically. He was still none the wiser as to what the entity's motives were that night. Had it simply remained still and silent, shrouded in the darkness of the branches, Jonathan was convinced that they would have walked right past it without ever realizing the creature was there.

Instead, the thing chose to confront Jonathan. Loudly and aggressively.

Could it, Richard asked, have been deliberately trying to instill fear? Jonathan and Richard both turned to Tobias, and his extensive knowledge of the Mothman phenomenon.

"In numerous cases, the type and extent of fear the experiencers feel goes above and beyond what might be considered a normal reaction to encountering something so far outside of the normal everyday experience," he told them. "For some of these people, there is a definite sense that the fear is actually being *imposed*... that this is being done deliberately by something intelligent."

To underscore his point, Tobias related examples in which the experiencers felt inexplicably terrified before they even realized that they were not alone. Does this mean that some kind of telepathy is involved? Could the sense of fear be so strong that it is virtually palpable?

Jonathan's encounter with Mothman has left lingering scars. Even now, several years after that night, there remains a sense of foreboding whenever Jonathan goes outside at night. Even something as simple as opening the doors to let the dogs out brings its own sense of anxiety.

"It makes me uncomfortable enough that I always want to come right back inside," Jonathan admitted. Most potent of all is when Jonathan walks their mom out to the car prior to her shift beginning. "What freaks me out is wondering just what that thing would have looked like if it wasn't dark."

Most Mothman sightings, including many of those from Point Pleasant, occurred at night. Daytime sightings are relatively rare. In Tobias's opinion, the majority of those can be attributed to misidentified birds.

After locking the house up tight, Jonathan went to his bedroom. He had just undergone a frightening, haunting experience. He wracked his memory, replaying the brief but intense experience over and over again. Jonathan guesstimated that the entity was seven or eight feet tall, though it was impossible to be sure.

Three or four minutes later, Jonathan's father went outside. He saw nothing out of the ordinary. The red-eyed entity, whatever it had been, had vanished.

As researchers who were delving into the phenomena of Mothman and high strangeness, Tobias and Richard have long been fascinated by potential links between winged humanoid entities and UFOs. Sometimes, they found that some people seem to be prone to experiencing such things, finding themselves encountering the uncanny multiple times during their lives. These incidents can take multiple forms. For example, the trifecta might be experiencing a haunting, encountering a UFO/UAP, and also sighting a strange creature or entity. There's an old adage that the more of an interest one shows toward the stranger things in life, the more likely they are to show an interest back. Could this be the case with Jonathan?

One night, several years after the Mothman incident, Jonathan and a friend saw a UFO while driving home. At first glance, they thought it might be a drone; however, the size of what looked like a giant upside-down metallic pear soon put that idea out of their heads. The closest analogy to the shape they both saw might be a commercial or scientific balloon, although such devices are designed to operate at high altitudes and are never flown low in the sky over residential neighborhoods. Jonathan estimated the object's height at around twenty feet above the trees.

(Of particular note, the Google Loon Internet project was responsible for a Kentucky UFO scare in 2012. The purpose of Loon was to test the feasibility of providing internet access to remote parts of the world, via balloon-mounted broadcasting technology. Unfortunately, the designers were left with egg on their faces when the balloon escaped and went renegade. Along the way came a slew of 911 calls from individuals who thought an alien invasion was imminent.[64])

---

64. Tillman, "UFO Sighting? No, Just Google's 'Rogue' Balloon."

The flying pear (for lack of a better term) was embedded with numerous windows. Diagonal lines of bright light pulsed across its surface in waves. The object moved slowly, almost languidly through the air without making a sound. Beneath the roofs of their houses, the rest of the neighborhood slept or watched television, completely oblivious to what was hovering just a few feet above their heads. It was like nothing Jonathan had ever seen before—or since.

Darting inside the house to bring out a family member to see for themselves what was hovering in the skies, Jonathan emerged again moments later to find the object gone.

During the interview, Jonathan came across as completely sincere. Neither Richard nor Tobias was picking up on any hint of deception or delusion. He clearly believed what he was saying. That led Richard to ask a question he often posed to eyewitnesses of incidents that have taken place in the 2010s onward: Why didn't Jonathan take out his phone and either take photographs or, better yet, video footage of the object?

"I've thought about that quite a bit ever since it happened," Jonathan reflected. "When I got out of the car, I didn't have my phone. At that point, I still thought it was going to be a weird balloon or a drone—not a UFO. That's when she [Jonathan's companion] got out and pulled me into the car. She freaked out. By the time I could have gotten a clear shot, it had gone. It disappeared very quickly."

To the best of the two researchers' knowledge, nobody else had come forward to report sighting the pear-shaped UFO in the vicinity. There are around fifteen hundred thousand people living in Rockford. If any of them saw the same thing as Jonathan and his companion had, they kept it to themselves.

Getting to their feet, Tobias and Richard followed Jonathan outside. They wanted to stand in the exact same spot in which Jonathan had his potential Mothman sighting. They were both guided to a spot about halfway along the walkway, which led up to the front door. The yard was well cared for. There were flowers, neatly trimmed trees, and a stone birdbath formed by a trio of seahorses. Farther to their right, behind the corner of the house, stood a much taller tree, its uppermost branches standing at least five times higher than the roof.

It was among those branches that the red-eyed entity perched, amidst the leafy canopy. Jonathan extended a finger and pointed at the approximate area. The entity would have been, at a minimum, fifteen feet above the ground. They approached the tree and inspected it carefully. There was only so much that could be gleaned after the fact, but the researchers began to run through potential explanations for what Jonathan saw. He had imbibed no alcohol or other mind-altering substances—indeed, Jonathan was underage to drink at the time—and he had never experienced hallucinations before that he was aware of. (Nor has he since.)

Tobias likes to point out that many sightings of winged entities turn out, when thoroughly investigated, to be misidentified birds. That was a possibility. Some owls, such as certain herons and birds such as the stygian owl, are known for having red eyes, which can glow in the dark when reflecting light sources. Could Jonathan possibly have seen one of them, its eyes glowing red when the headlights of a passing vehicle lit them up?

No cars were driving past when Jonathan was walking back to the house. Jonathan's mother had already driven off, and it makes sense that if the high beams of a car could be behind the sighting, then Jonathan *and his mother* should have seen the glowing red eyes as soon as she fired up the engine. Neither of them did. It was only as Jonathan was making his way back to the house in the near darkness of the poorly lit street that he caught sight of the eyes. There was little in the way of ambient light to be reflected.

Although it can sometimes be difficult to judge size and scale at night, it's not easy to believe that Jonathan could have misapprehended even a relatively large owl for having been a huge black creature with "dinner plate–sized eyes."

There's also the fact that, while several different types of owls are native to Illinois, the stygian variety is not one of them.

The "it was a bird" explanation also ignores the issue of the screeching, and the palpable sense of terror it instilled in Jonathan. Whatever it was he encountered that night, it seems to defy easy and conventional explanation.

Could it have been Mothman?

That's certainly a possibility. As they said their good-byes to Jonathan and made their way back to the car, Tobias told Richard that their next pair of interviewees encountered something every bit as confounding—right in the heart of the town itself.

CHAPTER 10
THE ROCKFORD INCIDENT

The next meeting took place in the early evening. It was still warm despite the sun being low in the western sky. Tobias pulled into a parking lot that bordered a slow-moving creek. Separating the creek from the parking lot was a muddy track that locals used to walk their dogs or, for the more adventurous, take motorbike rides, despite the clearly posted No Motor Vehicles sign. The dirt gave way to a tarmac trail a few feet off to the right.

Richard followed Tobias down a small embankment, no more than five feet high, which sloped down to the track. Just ahead of the pair was an ugly concrete bridge. It carried a constant stream of traffic.

The chaps were there to interview two ladies who claimed to have experienced something remarkable at that very spot. After a few minutes of waiting, during which Tobias and Richard took a slew of reference photographs, a car entered the lot behind them and parked next to theirs.

Barb and Shana were a mother and daughter whom Tobias had previously met while shooting *Unsolved Mysteries.* At the time of their encounter, they were living in an apartment just a stone's throw away from the creek. Barb took a moment to point it out. The building had a clear line of sight to where the small group was standing.

It was around 1:00 a.m. on a pleasant June night in 2022. After returning from a coffee run, both women were sitting in their car with the windows down to let in some cool air. They had parked outside the apartment and were chatting.

A loud screech shattered the stillness. It came from the direction of the creek. They turned to look. What they saw stunned both Shana and Barb to their core. A large shadowy form was crawling up the embankment. Its form was silhouetted against the cement backdrop of the bridge and was backlit by bright security lights from a building on the opposite side of the creek.

"This thing was pitch-black," recalled Shana, echoing Jonathan's description of the winged entity he had encountered in the tree outside their home. "It was hunched over and crawling on all fours."

As it kept crawling in their direction, more physical details slowly revealed themselves.

It was humanoid in shape; although it possessed arms and legs, no hands or feet were discernible. Then again, this sighting took place at night from a distance, so it's possible that digits were too small for the eyewitnesses to make out clearly ... either that, or the entity simply didn't have them.

The creek was dry at the time, as it usually is in the summer months. As they both watched, the figure spread a wide set of batlike wings and launched itself into the air, then proceeded to fly down the length of the tree line on the near side of the creek—the same side on which they, Richard and Tobias, were now standing.

Recalling that the creature was "very tall, with a huge wingspan," Shana and Barb noted that this entity lacked the distinctive glowing red eyes that are reported by many eyewitnesses of Mothman encounters. After a moment, the winged being disappeared among the trees.

How tall? Seven to eight feet, by their estimation. This also tracked with Jonathan's estimate.

Rather than flapping its wings like a bird, the airborne humanoid appeared to glide. It reminded Shana of a massive gargoyle. Now, Richard asked the difficult question.

"I'm assuming that both of you had phones with you?" They each nodded yes. "Didn't it occur to you to take photographs or video footage of this?"

"It was over too quickly," explained Shana.

"We were both shocked by the situation," Barb added.

"All we could think was, What is this? What's happening? What is it going to do?" Shana agreed.

What happened next?

"We got out of the car and followed the tree line down toward the park at the far end." Barb pointed off into the distance, along what was then the entity's flight path. Unfortunately—or perhaps fortunately, for who knows what might have happened if they had encountered the creature again—they had lost it.

Yet the night's high strangeness was far from over.

Returning to their apartment, the two women stood looking out toward the creek, talking excitedly about what had just taken place. The sky directly above the building with the security lights had begun to glow as a swirling amorphous form started to materialize. Barb describes it as being something akin to a giant owl in shape. It reminded Shana more of a stingray.

The strange glowing mass moved soundlessly through the air, passing about thirty feet above the street before returning to its point of origin. It repeated the process a second time, then promptly disappeared.

The experience was over in mere seconds. Shana was struck by the sheer smoothness of the anomaly as it slid through the sky and back. The whole thing had transpired without a sound.

Tobias and Richard were struck by the commonalities that this encounter had with Jonathan's sighting. The terrifying screech. The blacker-than-black form. Both incidents occurred late at night, when most people were asleep. They each happened in relatively urban areas, not out in the sticks. In neither case were reports made by other people in the vicinity.

Then there was the link with another aerial anomaly. In Jonathan's case, this was more of a classic flying saucer sighting. Shana and Barb saw something less traditional, but equally bizarre and perplexing.

A couple of months later, once again in the early hours of the morning, Shana was outside after returning home when she heard the shriek again. She wasted no time in checking out the creek, along with her mother, but this time, there was no entity to be seen. All was still and empty once the echoes of the shriek had faded away. Of the winged humanoid, there was never any sign—which prompts the question of what exactly had been the source of the cry.

There were differences as well as commonalities between Jonathan's sighting and that of Shana and Barb. In the latter case, neither eyewitness felt the paralyzing sense of fear and intimidation experienced by Jonathan. What they felt was a sense of deeply profound shock.

Although there were no reported sightings that same night, this vicinity has seen more than its fair share of winged humanoid encounters. Barb, aware of his status as a seasoned cryptid researcher, turns to Tobias and asks him what he thinks it is that she and her daughter saw on that balmy June night.

"Why don't they find any dead bodies? Why are there never any young ones found?" she wanted to know.

"If this truly is an objective phenomenon, then all the evidence points toward it not being a biological one," Tobias explained. "There are no signs of predation—of these entities eating anything. Nobody has ever found a lair or someplace where they might live. It is a mystery, plain and simple."

A mystery that defies easy explanation, Richard was coming to understand; something that Tobias has known for a long time.

"What about the glowing thing we saw afterward?" Barb pressed him.

"Within ufology circles, there has long been conjecture about the existence of atmospheric beasts. Entities that live in the atmosphere of our world, and don't come down to the surface. They stick to their atmospheric habitat, in the same way that oceanic creatures like whales don't usually crawl out of the sea onto dry land.

"Now, I'm not saying that what you saw *is* that," Tobias went on, "I'm just noting that there is a precedent for things that are similar to what you describe. Such reports are rare, and we don't tend to get many in the twenty-first century. They were more popular back in the 1960s and 1970s. But it's certainly a possibility."

The creatures to which Tobias was referring have been described as giant floating jellyfish or undulating wingless snakes that somehow contrive to twist their way through the skies. One such case was reported in a Crawfordsville, Indiana, newspaper on September 5, 1891, when a pair of ice delivery men claimed to have sighted an eighteen-foot-long, one-eyed serpent some three or four hundred feet above their heads. Pure white in color, the squirming beast was propelled by a series of fins running along the length of its body, which it swished in the way a fish would when moving through the water.[65]

According to an article published in the *Crawfordsville Journal*, the two men watched the creature for almost an hour and then fled. Obviously, there was no way for them to photograph the strange beast in 1891, and so we are left with only their word. If they were hoaxers, one wonders why they would have bothered to cook up such a story. What was in it for them?

Although the obvious skeptical explanation would be that they either made it all up or were possibly drunk at the time, this is discounted by supporting testimony from

65. Lighty, "The Crawfordsville Monster."

a local pastor, G. W. Switzer, who saw the same thing himself—two hours before the icemen's encounter. He estimated the creature was about the same size as they had reported (around sixteen to eighteen feet long and eight feet wide) and described the same twisting, squirming type of movement.

At the time, one potential explanation was cited as being a flock of birds, with white feathery undersides giving the creature its distinctive appearance. Supposedly the birds became navigationally challenged, leading them to circle over the town of Crawfordsville for approximately two hours before finally disappearing into the darkness of the predawn.

Other skeptics dismissed the sighting as having been nothing more than an uncrewed rogue balloon, released into the sky and allowed to drift away.

Either hypothesis offers up a potentially credible explanation, though one is forced to wonder how three grown men of presumably sound mind could so badly misidentify mundane objects or creatures. Could it instead have been one of the atmospheric dwellers to which Tobias is referring?

"Your experience of seeing the floating light is not common," Tobias told Barb and Shana. "I've never heard anything like that in connection with Mothman or the winged humanoid phenomenon."

He hypothesized that the reason neither of the two women felt any kind of fear was that the flying entity seemed to be completely unaware of their presence. In the case of the terror and angst experienced by Jonathan, the creature skulking in the tree was looking right at Jonathan. This means that there was a sense of connection between Jonathan and the entity, one which may have been a predator-prey dynamic. This was not so with Shana and Barb. Whatever it was they encountered did not so much as acknowledge their existence.

Could the presence or lack of awareness and interaction somehow be linked with the fear that some Mothman or winged humanoid witnesses deal with? It was an intriguing notion, and one that deserved further consideration. Tobias and Richard thanked their interviewees and said goodbye.

Their time in Chicago was at an end. The next leg on their journey would take them to a small town in Iowa, one said to have been terrorized by a bulletproof flying entity over several nights in 1903.

CHAPTER 11
THE SCIENCE OF MOTHMAN

In his capacity as a cryptozoologist and author, Ken Gerhard has crisscrossed the globe in search of legendary and fantastical creatures.

Growing up with a scientist for a father gave him valuable insight into the scientific method and helped shape the way young Ken viewed the world. His field of endeavor was forestry, which meant Ken accompanied his father on numerous outdoor trips and expeditions. He grew comfortable in the woods and the wilderness and developed a fascination for animals, which led him to collect snakes and salamanders among a menagerie of other critters.

Ken's mother was a travel agent and took her son to different countries whenever she traveled herself.

Couple that with Ken's boyhood love for monster movies such as *Godzilla* and *The Creature from the Black Lagoon,* and it's not difficult to see how he ended up where he did. Books taught him about such beasties as Bigfoot and the Loch Ness Monster, catching the youth at a formative age and becoming highly influential upon his life. The fifteen-year-old Ken traveled to Loch Ness with his father. The latter fished, while Ken attempted to capture footage of the infamous monster on his eight-millimeter movie camera. (Sadly, he never found the notoriously shy beastie.)

As he grew older, Ken searched for lake monsters in Canada; went on the trail of Bigfoot across North America; sailed the waters off the coast of Holland looking for giant sea serpents; and went Down Under to Australia in search of the Aboriginal legend, the Bunyip, to name just a few.

Not to mention, Mothman.

"This is basically a synthesis of all my passions," Ken explained. He was dealing with the vestiges of a summer cold when we sat down to speak with him. Sitting comfortably in his home office, Ken sported a jaunty black leather cowboy hat, which only served to buttress his colorful image. Behind him were row upon row of books and DVDs, almost all of them having a weird or monstrous theme.

Tobias and Ken were old friends, well acquainted with one another and their mutual bodies of work. Tobias explained Richard's status as a relative newcomer to the Mothman arena, providing some valuable context before the conversation began.

"My mother used to tell me all about the West Virginia Mothman when I was growing up. It was one of our favorite stories," Ken recalled fondly. He visited Point Pleasant for the first time in 2010. That opened the door to him researching what he soon realized was a much broader phenomenon of Mothman-like flying humanoid cases, and those of winged cryptids in general.

Ken had a wall-mounted map full of pushpins, each of which indicated a sighting of winged cryptids across the United States. Each pin was color coded to identify the type of anomaly that is reported. One color represented large avians, such as the Thunderbird, a supernatural birdlike entity Indigenous peoples believed to protect them from evil and darkness. Another color represented pterosaurs and pterodactyls. A third represented airborne entities that appear to be humanlike but possess wings—such as Mothman.

It didn't take long for the cryptozoologist to note that these sightings tended to appear in clusters, and that multiple different types were represented in the same cluster. In other words, a mix of flying anomalies are often reported in the same geographical region in approximately the same period.

"Do you have a hypothesis as to why the different types might gather like that?" Tobias asked.

"The more pragmatic hypothesis would be *interpretation*. People see something unknown that has wings. Some people perceive that to be Mothman. Others see it as a pterodactyl or a Thunderbird. It depends on their individual belief system and biases." Ken leaned back in his chair, thinking. "We're not talking about prolonged observation here. These are usually nothing more than fleeting glimpses. They often happen in emotionally charged situations. People are surprised, scared, or in a state of shock."

He made a valid point. Consider what happens when somebody is the victim of a violent crime, such as an assault or a mugging. They're typically not expecting to be attacked beforehand. Suddenly, the unexpected strikes. Their sympathetic nervous system kicks in. It's responsible for the fight-or-flight response that happens when jolts of adrenaline surge through the body, making the heart pound, the hands shake, and the pupils dilate. Every criminal investigator knows that such victims, through no fault of their own, are not always reliable historians. Their testimony must be considered carefully, filtered as it is through the emotional blinders that emerge when crisis strikes.

"Some of these sightings can be put down to simple misinterpretation of the unknown," Ken went on, "classifying it as best their mind can while it's caught in a difficult situation."

The second hypothesis Ken advanced was a more interesting one.

"On the other hand, we could be dealing with an entirely metaphysical or supernatural phenomenon, one that is somehow able to change its form and shape-shift ... something capable of looking like different things to different people.

"Imagine a phenomenon that could channel into the observer's unconscious mind and generate an image of whatever monster or creature is going to scare that person the most," he continued. "This goes back to the concept of the Cosmic Joker."

The authors referred to this concept briefly earlier in the book. It is the notion that much of the weirdness and high strangeness documented by researchers such as Charles Fort and John Keel may be explained by the presence of a mysterious intelligence that has quite the warped sense of humor.

To be clear, Ken was proposing a phenomenon that is drawn from the imagination of the observer, not something flesh and blood. In his view, Mothman wasn't *biologically* real (which is not to say that the phenomenon isn't real in any way, shape, or form). He pointed out that this creature, as described, could not have evolved from any currently known source, and it breaks the laws of physics as science understands them.

Whatever this thing was that people were seeing, it would have been impossible to shoot it down, drop a net on it, or, in most cases, physically touch it. There was probably a reason why nobody had collected a sample of Mothman feces. On the other hand, there *are* instances in which the winged figure appears to have left tracks or footprints.

"It's important to look at this phenomenon with a broad view, rather than cherry-picking or microfocusing on just a few aspects," Ken expounded. "Reports of physical contact are very rare in the scheme of flying humanoid encounters. I've interviewed dozens of witnesses. They definitely *see* these things, but I don't know that they're hearing them or smelling them. However, proponents of other paranormal entities such as ghosts and demons will talk about their having a physical presence. I personally have seen a door slam all by itself in a supposedly haunted house. We couldn't debunk it or explain it. That was an unexplained physical action taking place within our own physical reality. So, it's conceivable that the Mothman phenomenon behaves in a similar way … something that can manifest physically into our own realm, albeit only for brief moments of time."

Ken switched gears momentarily from the metaphysical to the cultural historical perspective. "This is a *very old phenomenon*. It's been with us since long before the Point Pleasant sightings of the 1960s. If you look at the beliefs and mythologies of different cultures all around the world, you will find traditions of anthropomorphic winged creatures everywhere. There are cave markings dating back to ancient France. Visual representations also come from Egypt, Sumeria, Assyria. Let's not forget the Harpies ..."

Harpies are probably the most recognizable winged humanoids, originating as they do in the mythology of ancient Greece. Carvings of harpies survive on tombs and burial markers dating back to that period. Homer preserved the harpy for posterity in his epic poem *The Odyssey.* The 1963 motion picture *Jason and the Argonauts* saw special effects maestro Ray Harryhausen bring the creatures to life using stop-motion techniques. They appear as horned, bat-winged creatures, with taloned legs and claws. Their sole purpose was to torment the blind man Phineas, snatching away his food whenever he sat down to eat.

Mythological harpies were generally given female characteristics, though there are also instances of male variants. Sometimes, they could be fair of face, beautiful, or handsome; at other times, they were depicted with a monstrous visage. Their behavior tended to range from being mischievous to outright cruel. Today, it is not uncommon to refer to a woman as a harpy, suggesting that she is shrewish and unpleasant in nature. One never hears men referred to by the same name. The ancient patriarchy strikes again.

"There are two ways we can look at this," Ken went on. "If this is a purely cultural phenomenon, then it's clearly one that we've evolved with over tens of thousands of years. If that's the case, then *of course* we're still seeing winged humanoid-type creatures in our skies. Equally, if this is supernatural, a metaphysical construct, then it would also make sense that humans have been experiencing this for millennia. Of course, people didn't just start randomly seeing these things starting a few years back."

One aspect of this whole thing that both Tobias and Richard were interested in pursuing further was the possibility of a connection between Mothman-centric high strangeness and other subsets of the paranormal, such as ghosts, UFOs, and psychic powers. It was something to be borne in mind as the search for Mothman progressed.

The topic of conversation turned back to the clustering of Mothman-like sightings in certain key areas. What could be behind that? Ken observed that many of them occurred near military bases and defense installations. The same appeared to be true of civilian airports.

"Even old Spring-Heeled Jack made an appearance at an army base, back in the day," the cryptozoologist chuckled.

Spring-Heeled Jack was a British phenomenon that set national interest afire in the Victorian age. Much like Mothman, he was humanoid in nature, but descriptions of him (if Jack was indeed a male, or even human at all) varied widely. Some depictions showed Spring-Heeled Jack as a dastardly human man—others, as more of an inhuman beast. Some newspaper and magazine cartoonists of the era portrayed him with a distinctly Satanic mien, complete with a flowing black cape, forked beard, and lascivious leer.

In all cases, however, he was said to have leaped high through the air with a single bound, sometimes clearing houses and other buildings with apparently minimal effort. From the 1830s all the way through the early 1900s, this bizarre bounding character was seen across the English landscape, from towns to major cities to the countryside.

There were claims that Spring-Heeled Jack assaulted women, though just how much of this was simply hysteria and penny dreadful–style reporting should be considered carefully. Tellingly, some eyewitness accounts told of Spring-Heeled Jack having flaming eyes—a clear parallel with the Mothman and its glowing red eyes.[66]

One key difference was that Jack left footprints behind him, though it was impossible to follow any sort of trail. He struck like lightning out of a clear blue sky, pounced on his victim—often a "defenseless" lady—then bounded back into the air and made his escape. Sufficient consternation arose from some incidents that the government became involved, particularly when the attacks took place in London. Supposedly, Sir Arthur Wellesley—the Duke of Wellington and victor of Waterloo—organized patrols to track down the elusive creature. They were unsuccessful.

---

66. Castelow, "Spring Heeled Jack."

The military base incidents to which Ken referred involved encounters at Aldershot and Colchester Garrison. Today, Aldershot is the home of the British Army. Its airborne forces are based there. Colchester is better known for housing Britain's military prison, or "glasshouse." In 1877 and 1878, Spring-Heeled Jack descended from the skies and attacked sentries at Aldershot, for no apparent reason. The sentries opened fire, but either missed their assailant completely, or he/it was somehow impervious to their bullets.[67] In the Colchester incident, a determined soldier is said to have lunged at Spring-Heeled Jack with his rifle bayonet and stabbed him/it in the leg. It was only then that a more prosaic explanation came to light for the latter case, at least; like the consummate Scooby-Doo villain, Spring-Heeled Jack was unmasked and turned out to be a junior officer who was clowning around.[68]

Leaving aside the tomfoolery of young military men who ought to have known better, the parallels between Spring-Heeled Jack and Mothman were striking. Both were humanoids that were seen to soar through the air, often inciting great fear in those who encountered them. Both were described as having had glowing or burning eyes. A key difference is that Spring-Heeled Jack physically attacked individuals, whereas Mothman seems to spurn physical

---

67. As reported in the *Times*, April 28, 1877.

68. Bell, "The Legend of the Spring-Heeled Jack."

contact—if, that is, Mothman is even capable of interacting with our material world.

Another similarity was the wide variance in physical descriptions. "Some witnesses put Mothman and fellow winged humanoids at about six feet tall," Ken said, checking characteristics off on his fingers. "Others say twelve feet tall. Sometimes there are horns. Sometimes there are glowing eyes. Back in the 1950s, there was a flap in which the airborne figures were biomechanical, with aluminum-looking wings and electronic control panels."

It was almost as if the phenomenon has been changing over time, morphing its appearance to keep pace with the development of the society that experiences it. This brought to mind the halcyon days of the UFO phenomenon. Initially, those who claimed to have contact with flying saucers said that, according to the craft's inhabitants, they came from bases on our moon, or nearer planets such as Mars and Venus. Once humanity sent probes to those planets and found them to be not just unoccupied, but, in the case of Venus, actively hostile to anything we might consider to be sentient life, the story changed. Now the extraterrestrials claimed to come from farther out in the solar system.

When technology afforded humanity the opportunity to explore those, and they turned out to be similarly devoid of life, the occupants of the flying saucers now claimed to come from other stars. As humanity's scope expanded farther, the phenomenon withdrew, morphed, or did both, always remaining maddeningly just out of reach.

This led Richard and Tobias to a question they simply haven't been able to answer to this point. In the twenty-first century, when there were more cameras installed per capita than at any other time in history; when most people in the Western world were walking around with a high-definition camera in their pocket, incorporated into their ever-present cell phone—where was the compelling video footage (or even convincing single-frame photographs) of Mothman? Statistically, there should have been at least *something* compelling captured on camera by now.

Ken expelled a long breath as he considered the question.

"You could make that argument about pretty much any cryptid. Like many people, I'm not really impressed by photographic evidence these days. If we had a video, would we even believe it was real? Most people would dismiss it as being AI-generated or digitally doctored in some way. Fakery is everywhere. Video evidence is losing its importance."

He was absolutely right about that. Of the many ostensibly anomalous video clips that air on "send us your paranormal footage" TV shows, few people have found much of the content to be compelling in terms of evidence. It has entertainment value, yes, but evidence? Likely not.

What did Ken Gerhard make of the "Mothman as harbinger of tragedy" narrative?

"There *could* be a level of manifestation that we are all capable of achieving, either collectively in a Jungian way or, alternatively, as individuals. Who knows what people can achieve? A neuroscientist once told me that even after decades of research, science still has virtually no idea of how the human mind actually works. Yes, it's known that memories are formed because of the bonding of various proteins, but when it comes to the big picture, we're still quite ignorant."

In other words, we may still be incapable of truly understanding the true scope and nature of the Mothman phenomenon at this point in time. Our understanding of the mind, of physics, is too infantile. Our frame of reference is simply too limited.

"When people encounter Mothman, they're typically stunned by the experience," Tobias pointed out. "This is usually given as the reason for them not having taken a photograph. Speaking as a man who can't even get his phone out fast enough to take a picture of our dogs when they're doing something cute, I can relate to that."

He also noted that cell phone cameras are not designed to take photographs or video of fast-moving aerial objects. Apart from a select few models at the high end of the range, such cameras are optimized for taking selfies and portraits ... not Mothman.

"Being surprised and overwhelmed with emotion could definitely be a factor as to why eyewitnesses don't seem to produce pictures of their Mothman sightings," Ken conceded. "But there are surveillance cameras and Ring doorbells *everywhere*. You would think that someone, somewhere, would have gotten some interesting potential Mothman footage by now, even if only by happenstance."

So, what *did* Ken consider the most compelling evidence pertaining to the existence of Mothman?

"It would have to be exclusively eyewitness testimony," he concluded, steepling his fingers. "As is the case with most cryptids, the body of reports is largely anecdotal... a combination of credible modern witnesses, whom we can interview, and then folklore and legend also factor into it. Multiple witness encounters are the crème de la crème of course."

It made sense that those instances in which Mothman had been spotted by more than one person at the same time carried more weight than cases centered on a lone individual. Corroboration can be a powerful thing, which was why Ken found the Scarberry/Mallette sighting to be one of the strongest pieces of evidence out there. (This story, part of the Point Pleasant Mothman flap, was detailed earlier in the book.) While the account is debated and some of the details are contested to this day, there was a definite consistency to the testimony. There could be no doubting that everybody involved experienced... *something*.

"I consider it to be more credible when the witness or witnesses call the police or involve officials in some capacity to make an official report," added the cryptozoologist.

"Suddenly your name could be in the newspapers and on TV. Why would they want to give out false information publicly?"

The obvious answer is the old saying that "there's no such thing as bad publicity." This statement plainly is not true. Although the public has grown significantly more tolerant of beliefs that would once have seemed outlandish, back in the 1960s and 1970s, telling the world that you saw Mothman flying around could have been a surefire way to get yourself branded a crackpot... hardly the kind of publicity that somebody would want their family or employer to get wind of.

Ever since Mothman first made its way onto the global stage, there have been enthusiasts. Many people find the creature's silhouette instantly recognizable.

"In your opinion, how prevalent is preexisting bias?" Tobias asked. "Are witnesses often interested in Mothman prior to the encounter, or are many of them relatively ignorant on the subject?"

"It runs the gamut, but the majority of witnesses I know about seem to have Mothman on their radar somehow. Maybe they saw the movie, saw the show, or read a book. So, yes, there's a degree of confirmation bias, and it's exacerbated by the popularity of monsters in our everyday culture, thanks to videos, movies, podcasts, and so on. Cryptids have so much more exposure today than they did when I was growing up. That's why I find some of the more recent encounters less credible."

All three researchers—Ken, Tobias, and Richard—were children of the 1970s and 1980s. When they were growing up, they got their weirdness fix from the occasional TV show such as *In Search Of ...* in which Leonard Nimoy ditched the Mr. Spock ears and profiled cases of the paranormal and outright bizarre. Now, there are entire streaming channels dedicated to such content, and thousands of creators vying to tell similar stories. In contemporary Western culture, one would have to work hard to *not* be aware of Mothman.

"Just look at the Mothman Festival," Ken went on. "Over fifteen thousand people went to the 2023 event in Point Pleasant.[69] So, yes, people *are* more biased today than they have been in the past. They're absolutely being culturally influenced."

Richard asked where the majority of flying humanoid cases are currently occurring. Ken referred to his case files and revealed that there have been multiple potential Mothman sightings taking place in Northern Mexico and the Southwestern United States, particularly West Texas, New Mexico, Arizona, and Southern California. These were all relatively sparse, desolate locations, arid and desertlike. They often also have a plethora of UFO sightings in conjunction with those of flying humanoids.

---

69. Vaughn, "Organizers: Over 15,000 Attend Mothman Festival."

Ken had noticed a fascinating correlation between Mothman sightings and those of another unusual cryptid: the Dogman. This enormous creature, which is said to have the body of a human and the head of a dog, supposedly prowls the wilds of northern Michigan. Reports went back to the late nineteenth century.

"I'm becoming convinced that the Dogman phenomenon is very close, if not almost identical, to that of the flying humanoid/Mothman," he revealed. "You have anthropomorphic creatures that are both a biologic impossibility, with blended characteristics of a human and some type of animal. Both can appear hyperaggressive chasing cars and witnesses at high speeds. Both have big, red glowing eyes. There are strange smells ... misty, green, amorphous hazes.

"Just like Mothman, the physical descriptions of Dogman are all over the map, ranging in size from four to seven feet tall, with different types of head. Both entities seem to feed off our fear, taking the form of our worst nightmares and terrifying people. They chase, but never physically make contact. Although there's occasional damage to cars and other vehicles, there are no direct human attacks, wounds, or physical trauma to human beings."

"This sounds a lot like John Keel's Ultraterrestrial hypothesis," observed Tobias, "where there's a single root phenomenon that may be at work ... one that takes many different forms. How much do you think the Ultraterrestrial hypothesis explains? Does it go beyond Mothman and Dogman to also explain Bigfoot, UFOs, and ghosts?"

"Keel was a great researcher, and I have much respect for him in a number of ways," Ken said, choosing his words carefully. "He also embellished some things and even outright made some things up. That being said, he was definitely on to something. He interviewed a lot of eyewitnesses and generated a lot of reports. He did identify this phenomenon, which seems to be constantly reinventing itself for different individuals."

Turning back to Mothman, Ken noted that both it and the Dogman seemed to represent the predator archetype. Their behavior, while not tangibly dangerous, does sometimes evoke that of a hunter stalking and harassing its prey.

"These monster archetypes always have human characteristics combined with those of creatures that are dangerous to humans," he added. "It seems like humans invent these things to be the scariest, most hazardous things in relation to us. I've often wondered whether this phenomenon is a form of supercharged pareidolia, with people seeing something that isn't *really* objectively there, but is instead being created by their minds. Perhaps these monsters are manifestations created by our own minds, an extension of fears our prehuman ancestors had when they lived in more dangerous times."

This was a fascinating concept. Could the Mothman in fact be nothing more than primal fear given tangible form? It was a concept worthy of further investigation.

CHAPTER 12

THE VAN METER VISITORS

Although Van Meter, Iowa, is designated as a city, it's really more of a town. It retains that homely small-town air that permeates so many settlements throughout the American heartland. Located some twenty miles west of Des Moines, around fifteen hundred people call Van Meter home, but it feels smaller somehow, more intimate. There's a post office, churches, a library, and that staple of the Midwest, a Casey's gas station.

The people are friendly and welcoming to visitors, as Tobias and Richard discovered when they arrived on Saturday, September 28. The date wasn't random. They'd chosen this particular weekend because of the Van Meter Visitor Festival: a celebration of one very strange week in 1903 that has forever placed the town in the history books.

Tobias had driven west from his home in Wisconsin. Richard came east from Colorado. We were accompanied by fellow investigator and native Iowan Sarah, rolling into Van Meter on a hot and sunny Saturday morning. One of the local bar-and-grill joints had a large sign that proudly declared Winged Creature Spotted Here.

If local lore and legend had a basis in truth, then the avian entity in question might tie in with our search for Mothman. Much of what is known about this weird episode comes courtesy of researchers Chad Lewis, David Weatherly, Kevin Lee Nelson, and Noah Voss, all of whom delved into the mystery and detailed their findings in several different publications. (The authors highly recommend reading their books, which can be found cited at the back of this one.)

Van Meter was a little over twenty-five years old when the winged monstrosity came to town in late September of 1903. Its arrival was heralded by a bright ball of light seen atop a building. There was no sign of an actual entity at that time. That would come shortly afterward. Yet with hindsight, it seems clear that this was when the phenomenon began: Monday, September 28. The eyewitness was a purveyor of farm equipment named Ulysses Griffith.

At first, Griffith thought the light might have indicated a burglary in progress. He slowed down to take a closer look, but before he could make anything out, the light disappeared, before instantaneously reappearing on a nearby rooftop. Then, in a flash, it was gone, leaving no hint of its origin behind.

The next night, a bright light woke Van Meter's resident physician, Dr. Alcott. The source was hard for him to credit: a bat-winged humanoid creature, which emitted the light beam from its distinctly nonhuman head. The good doctor reacted in a typically Iowan way to being confronted with what appeared to be a monster: he grabbed the nearest firearm, a pistol, and opened fire.

Either Dr. Alcott was a terrible shot, or the beast was bulletproof. None of the shots he fired seemed to have any appreciable effect. Recognizing that discretion is sometimes the better part of valor, Alcott wisely chose to retreat.

For years afterward, skeptics would dismiss the entire Van Meter Visitor phenomenon as being nothing more than delusion or a hoax. Such things could not exist, the line of reasoning went, and therefore it did not. The eyewitnesses must have been drunk, lying, or at best gravely mistaken. Yet Alcott was a professional man, a doctor. As such, he should have been a skilled observer, trained in the scientific method. He had no reason to lie; indeed, telling monster stories would only serve to bring his credibility into question, something that in no way benefits a member of the medical profession.

At any rate, Dr. Alcott barricaded himself in his office, guns close at hand, and refused to come out until morning.

Apparently nocturnal by nature, the creature returned on night three/morning four. This time it was spotted by bank cashier Peter "Clarence" Dunn, who was pulling overnight duty as a security guard at the bank. When a blinding beam of light shone in through the bank window, Dunn opened up on it with his shotgun—without bothering to open the window first. Either he was just as bad a shot as Dr. Alcott, or the Visitor was indeed impervious to buckshot, because no carcass was discovered in the street outside. No sign that the beast had been injured at all... yet in a surprising turn of events, in the mud behind the bank, there were allegedly footprints. Mr. Dunn was said to have taken plaster casts. They had three toe-like protuberances. This is a rarity among flying humanoid cases, and although the Dufield Pond sighting that we investigated may also have left a single footprint behind, it was more humanlike in appearance.

Unfortunately, the whereabouts of the plaster casts have long since been lost to history.

The following evening brought another encounter, this time with hardware store owner O. V. White. White slept in a room above his property. He woke up in the early hours of the morning to a strange sound, something akin to two metal files or rasps being ground together. The sleepy Iowan went over to the window. Less than fifteen feet away sat the infamous Visitor, squatting on the cross-beams of a telephone pole.

The creature was so still that White assumed it was sleeping. (Could the rasping sound have been the entity breathing or snoring, researcher Chad Lewis speculates.)

Like his predecessors, O. V. White wasted no time in arming himself. According to the newspaper account, the storekeeper allowed a few moments for his eyes to adjust to the darkness, then made careful aim at the winged being and took a potshot at the Visitor from very short range. The creature awoke from its slumber and looked right at its assailant, shining the intense beam of light directly at him.

In an unexpected turn of events, the Visitor unleashed a foul odor. White would later claim that the stench affected his memory, inducing amnesia where the rest of that night was concerned. This effect was reminiscent of the missing time phenomenon reported by numerous UFO abductees, who remember little or nothing for hours or sometimes days after coming into contact with UFOs. Could something similar have been at work in Van Meter on that October night?

If he was expecting to have made a kill shot, O. V. White was to be disappointed. Once again, the entity seemed to just shrug the gunfire off as being little more than an irritant. Yet the gunshot did serve to wake up other citizens of

Van Meter, who came out of their homes and businesses to see the Visitor descending the telephone pole rapidly.

Once it reached the ground, the creature stood erect on two legs and began to rather bizarrely hop like a kangaroo. Then it spread its wings to their full extension. At that moment came the piercing shriek of the mail train. This seemed to spook the Visitor, which returned to all fours and began to lope away from the speeding locomotive. Shortly afterward it launched itself into the air and flew off toward the city limits.

Speaking to their neighbors the following day, White and his fellow experiencers' collective description of the creature matched that of Messrs. Alcott and Dunn. Now that the people of Van Meter accepted the reality of their peculiar visitor, the question remained: Where did the creature make its lair?

The answer appeared to lie on the outskirts of town. Near a brick and tile factory was a disused coal mine shaft. The factory was still operational, and workers there reported hearing odd noises coming from within the depths of the earth. Further investigation revealed that not only was the bat-winged Visitor lurking within the mine shaft, but also that it was accompanied by a smaller version of itself—perhaps its offspring?

The people of Van Meter armed themselves and staked out the mine entrance overnight on the fifth night since the creature arrived in town. The two flying beasts returned at sunrise and flew into a barrage of gunfire that would have done an infantry platoon proud. Although apparently unharmed by the fusillade, the creatures were at least driven farther back into the mine.

The gun-toting workers quickly reasoned that if they lacked the firepower to kill the creatures, the next best thing would be to detain them. To this end, the mine entrance was swiftly boarded up. One version of the story has the workers dynamiting the mine to exterminate the creatures.

What happened next is anybody's guess. The mine entrance apparently remained undisturbed. The flying creature(s) never returned to the streets of Van Meter. Perhaps they still reside within the mine, hibernating or lying dormant. Or maybe there was another way out of there, and they took off for pastures new, putting as much distance as possible between themselves and the small Iowa town that liked to shoot them up.

There have been a handful of sightings that may or may not have been the Visitor over the decades. Strange flying creatures, some of which are humanoids and others that appear to be more avian in appearance, have been reported not just in the vicinity of Van Meter, but also across the United States and beyond.

At the town visitor center (no pun intended), Tobias and Richard enjoyed a day of lectures and presentations.

All the topics were suitably esoteric and interesting. The people of Van Meter have come to embrace their monster wholeheartedly. They're affectionate and proud of it. The flying beastie can be found emblazoned on T-shirts, bags, and posters.

To explore the potential connection with Mothman and other flying humanoids/cryptids in more detail, Tobias and Richard sat down with author and researcher Chad Lewis to ask him about it directly. They tracked Chad down at his vendor table, where he was signing books and chatting with attendees. As the end of the day drew near, the crowds were starting to thin out, affording them enough time to have a conversation about the Van Meter Visitor.

Tobias was intrigued by the origins of the Visitor story—the manner by which it came to the attention of the world at large.

"As far as we're aware, the news first broke in one main newspaper report at the time of the sightings," Chad began. "That was followed by a few reactions, letters written in by readers who either believed the story was genuine or thought it was a bunch of hogwash. There's really only one long main article, which was written by the postmaster H. H. Phillips."

Phillips was the guy who gave the world the story of the ball of light ushering in the arrival of the Visitor on the first night. Mobile balls of light (or simply strange lights) are often reported in incidences of high strangeness.

"Do these lights signify that something *other* is entering our reality?" Richard wondered aloud. "Like the light spilling out of a doorway when you open the door to your home at night?"

Tobias pointed out that, per John Keel's Ultraterrestrial hypothesis, the balls of light represented the natural state of the Ultraterrestrials—their normal, resting form, stripped of any illusion or deception.

"When humans see something else," he continued, "like, say, Mothman or the Van Meter Visitor, that's only because the Ultraterrestrial is manipulating your perception. Making you see what it wants you to see."

Unbidden, Richard's mind flashed to the climax of the movie *Ghostbusters*, in which the titular quartet square off against the evil god named Gozer. They're invited to "choose the form of the destructor." After Bill Murray's Peter Venkman cautions the Ghostbusters to keep their minds blank and avoid thinking about anything, it is the hapless imagination of Dan Aykroyd's Dr. Raymond Stantz that conjures up the image of a giant Stay-Puft Marshmallow Man. Are those who see Mothman or the Visitor unwittingly doing the same thing—subconsciously choosing to see the form of a winged monster? Or is the intelligence that may lie at the heart of the phenomenon selecting that form for them, and projecting it into their perceptions?

"By the fourth night, the townspeople were asked to light their lanterns and turn on all the lights in the town,"

Chad continued. The goal was presumably to make Van Meter as brightly lit as possible. This also makes it less likely that the residents were misidentifying something they only caught glimpses of in the shadows.

On the first night—September 28, 1903—the moon was in its first quarter; 57.8 percent of it was visible, shining a respectable amount of light down onto Earth. By October 1, the moon entered its waxing gibbous phase, with 82.62 percent of its surface available to reflect light onto the streets of Van Meter. The final night saw that percentage rise to 89.25 percent. Add to that the glare put out by every artificial light source that could be mustered in the town, and even though the skies were somewhat cloudy, there was likely no problem with illumination.

"It's important to note that this particular period of US history was not necessarily known for its high quality of journalism," Tobias pointed out tactfully. "A number of small towns were coming up with monster stories, often just as a way of bringing in more people—and therefore commerce."

When Chad first came to Van Meter, the owner of the land on which the old brick factory lay told the researcher that, when he and his friends played outdoors during their childhoods (the man was then in his seventies), they always hesitated to go near the ruin and the mine entrance. Despite the fact that a crumbling factory would seem like the perfect playground for a gaggle of adventurous young boys, they always felt as if something was strange about it…something was just off.

"This was a down-to-Earth Iowa farmer," Lewis mused. "A real salt-of-the-earth guy. Not the kind of person that is prone to flights of fancy. He said that he hated that area."

This would have been the 1940s, so Richard wondered aloud whether the boy had grown up hearing fanciful tales of the Van Meter Visitor.

"No," Chad shook his head. "In fact, when we first came out here to start our research, many of the local residents were unaware of the monster story."

Weirder still, after the initial reportage of the episode in 1903, there seemed to be no further mentions of it in the press until the twenty-first century. It's almost as if the town of Van Meter developed a sudden case of collective amnesia.

That definitely no longer remains true today. The Van Meter Visitor Festival is living proof of that. So is Point Pleasant's Mothman Festival, something that Richard and Tobias will soon get to experience for themselves.

Never one to shirk the difficult questions, Tobias asked Chad whether this could simply be a made-up story, an attempt to lure visitors to a town that otherwise has few attractions. Alternatively, was this simply something they did for fun—a prank cooked up for a lark?

"You can never rule out that possibility," the researcher admitted, "but a couple of things do stand out. The first is that all the witnesses whose names appear in the newspaper were prominent members of the community. The banker. The doctor. The businessman. Even a future mayor. Back then, your reputation was all you had. I don't know of many people who'd want to put their lifesavings in the bank that's managed by a guy who's blasting out its windows, shooting at monsters!"

He made a valid point.

"If they *were* hoaxers then they marketed this thing brilliantly," Chad laughed, "because they wrote in to the newspapers and told people not to come and visit Van Meter—that there was nothing here to see! Of course, that could also have been reverse psychology, hoping that people *would* come. If that was their intent, it never worked. The town never capitalized on the Visitor until this festival started."

"Reputation *is* important," Tobias conceded, "but maybe not as important as town revenue. Those reputations probably wouldn't have been damaged if everybody in Van Meter knew that this was all being done just for fun."

Chad wasn't buying it.

"If H. H. Phillips made all this up, then he was decades ahead of the curve with regard to strange phenomena and ufology." He began ticking points off on his fingers. "There was the period of missing time; screen memories; creatures on which weaponry of the time had no effect."

"Or he had one heck of an imagination," rebutted Tobias.

"A lot of people do say that this was all just the work of a brilliant storyteller," acknowledged Chad. When he first arrived in Van Meter, he spoke with a resident of the town who had been a boy when the actual experiencers of 1903 were old men. An old man himself at the time of his interview with Chad, and now sadly passed, he told the inquisitive researcher, "Those men ran the town. If they said there was a monster—then we all believed *there was a monster*." Their word was law.

If this was a hoax designed to put Van Meter on the map, and thereby generate profit, then it failed. Until fairly recently, the town was all but dying. The expansion of Des Moines has brought some new residents to town, but in 2025 it remains a relatively small and homely place.

For decades after the departure of the Visitor and its smaller companion, strange things continued to happen in and around the vicinity of Van Meter.

"In the 1980s, a man was walking his dog along the gravel road that runs parallel with the old brick and tile factory," Chad said. "A massive bat, roughly five feet in size, flew overhead. The dog cowered in fear. The man never walked his dog along that road ever again."

The largest bat in the United States is the greater mastiff bat. They tend to have a wingspan of less than two feet—less than half the size of the creature in question. An oversized bat is one of the possibilities that has been put forward to explain the mystery of the Van Meter Visitor, yet even if that hypothesis is correct, it fails to explain why the creature was apparently impervious to gunfire.

On the opposite side of Des Moines to Van Meter is the city of Colfax. It, too, is relatively small, with around twenty-two hundred residents. The two cities are forty-five miles apart. In 2006, Chad tells us, a pastor got the shock of his life when he saw what appeared to be a dragon flying through the skies above Colfax. An internet search revealed an artist's rendering of the Van Meter Visitor. The pastor recognized it immediately. So far as he was concerned, the dragon and the Visitor were one and the same.

If this really was the case, 103 years separated the two encounters with a very strange winged beast that soared through the Iowa skies. It may lack the glowing red eyes that often characterize Mothman sightings, but there are enough commonalities to give us pause for thought.

Just how far did this phenomenon extend?

Once their conversation was over, it was time for Tobias and Richard to follow in the Van Meter Visitor's footsteps … well, its glide path, more accurately.

Chad Lewis was an excellent tour guide, engaging and erudite. He told the story of the Visitor very capably, and the fact that he did it while walking in the footsteps of those who actually lived it only served to give the tale greater impact. Tobias and Richard were convinced that the single best way to hear this tale was from Chad himself, while roaming the streets of Van Meter.

The bank was still there, albeit considerably modernized since 1903. It was set up with a desk and shelving, the sort of place in which one could sit down and do some writing. Richard and Tobias took two steps inside the vault itself, and it required little imagination for either of them to picture a beam of intense light piercing the street-facing window some 121 years prior.

Chad pointed out each site of significance within the town before leading them out along a country lane that bordered pastureland. This was the same gravel track on which the dog walker had his frightening encounter with the massive bat. It was there, off in the distance and silhouetted against a copse of trees, that Tobias and Richard first set eyes on the dilapidated brick ruin that was once Van Meter's brick factory. It was now a pale shadow of its former self, little more than a crumbling wreck of a structure. In its heyday, however, the factory was a powerhouse, churning out a steady stream of bricks to be used by the construction industry.

It warmed Richard's dank, dark, ghost hunter's heart to learn that the site of the brick factory was also reputed to be haunted.

Although it was all but impossible to pick out with the naked eye, the grassed-over entrance to the mine still lay adjacent to it. It is most likely there that the key to the riddle of the Van Meter mystery can be found. Excavating the mine itself might yield the remains of the Visitor and/or its smaller companion, thereby settling the issue for all time. Unfortunately, it would take more time, money, and effort than anybody has thus far been willing to invest, and so the issue remained unresolved.

It should also be noted that, even were the mine to be exhaustively explored, that doesn't necessarily mean anything remains to be found in its subterranean depths. The creatures could be long gone. It's possible that, if they are either closely or even peripherally related to the Mothman phenomenon, then the Visitors may have been capable of simply phasing into and out of our reality in the same way that Mothman has been said to do.

All of which prompts the question: *Was there* indeed a connection between the Van Meter Visitor and Mothman? The Visitor had a horned, nonhuman head. The horn seemed to be a discharge point for the light beam it emitted. If descriptions are to be believed, the Visitor lacked the glowing red eyes we've heard so much about. The bat wings, however, did track with some other eyewitness accounts of flying humanoids.

To investigate this potential link further, Richard and Tobias were hitting the road again. This time, they were heading to what is unquestionably the epicenter of the Mothman phenomenon: Point Pleasant, West Virginia.

CHAPTER 13
LOVE 'EM OR HATE 'EM

Investigator Richard Estep was indisposed during the interview with writer Richard Hatem, so Tobias conducted it solo. Regardless, to avoid confusion, the interviewee will be referred to by either his full or last name.

Richard Hatem is the screenwriter responsible for the 2002 paranormal thriller *The Mothman Prophecies*.

Growing up as a teenager in the 1980s, Hatem wasn't exposed to the Mothman legend, although he did have an interest in more well-known monsters like Bigfoot and Nessie. It wasn't until April of 1997 that he would find himself absorbed by Keel's work on the phenomenon. Hatem had just read Patrick Harpur's *Daimonic Reality* a few months before when he first stumbled upon a copy of *The*

Mothman Prophecies in a West Hollywood bookstore. He was immediately enthralled by its contents.

"I found the book and I read the back of it, and I'm like, 'Oh my God, this sounds amazing. Okay, I've never read it. Now's the time. I'm going to read about the Mothman,'" Hatem told Tobias. "Then I started reading this book I'd never read, and was instantly drawn in by John Keel's voice, completely fascinated by the book. There are a handful of times when you're reading the book and just the top of your head comes off. And I was like, 'I don't know what I'm going to do, but I got to do something with this book.'"

By this time, Hatem was already writing screenplays, so he immediately contacted his agent about centering a project around Keel's work.

Excited, he pitched his idea, inquiring about acquiring the rights to *The Mothman Prophecies*. A couple of months later, Hatem's agent reminded him of his pitch about the "haunted town." The agent's inadvertent phrasing struck a chord with Hatem, who is fascinated by the breadth of phenomena present in Keel's book.

"What I loved about *The Mothman Prophecies* book was that it was not relentlessly about the Mothman. It was about this whole thing," Hatem said. "And look, John Keel's whole approach to this is weird. The MiB are weird. The UFOs are weird. People's interactions with the phenomenon are weird. It's all high strangeness, and the highest of all strangeness is sort of our mascot Mothman, this other thing that no one sees."

He continued, adding, "I love the fact that, if you will, Mothman sort of hovered over his own book, never really occupying it, but simply lingering in the shadows, representing everything that was unknown. I mean, it's a perfect metaphor. It's a little funny. It's a little weird. It's a little scary. It doesn't make any sense. There's sort of an explanation, but it doesn't really fit. It's like if Mothman didn't exist, you'd have to invent him to represent the archetypal encounter with the anomalous."

After that, while working on the screenplay, Keel's work seemed to follow Hatem wherever he went, until he'd read both *Operation Trojan Horse* and *The Eighth Tower*.

The speculative nature of these works appealed to Hatem, who appreciated Keel's approach to Forteana.

"It wasn't that sort of shrill, 'I have hard evidence and I'm going to expose the government.'" Hatem explained. "But it was a philosophy. Keel's philosophy of course was, 'Hey, the weirder it is, the realer it is.' And that was what appealed to me. And so that's what I tried to sort of hold on to when I wrote the movie."

Hatem's feelings on Keel were only reinforced once he had occasion to meet the researcher in person.

"He was the nicest guy," Hatem said of Keel. "I met him a couple of times, and he was always extremely friendly, self-effacing, happy to talk about whatever, but again, never banging the drum for anything. In my mind, he was going to be a guru. He was going to be like, let me tell you, here is the secret of the pyramids. And instead, he just wanted to tell funny stories about the people he knew and refused to take any of it seriously. He was just like, 'Look, I've been in three different houses with poltergeists while the poltergeist outbreak was happening. I've been here; I've been there. I've seen this. I've seen that. Your guess is as good as mine.'"

"That sounds really endearing, honestly, considering how many people take the opposite approach and try to proselytize every time they talk about this stuff," Tobias interjected.

But writing the movie didn't come easy, mostly because of how resistant the book's narrative is to translation.

Hatem knew that he couldn't adapt the work directly and make a palatable film, so instead, he aimed to capture the feel of the phenomenon.

"For me it was like, well, the tension of the book is that he's losing his mind, but I'm like, in a movie, what you want is a triangle. So, it's almost like a love triangle. It's like [the protagonist is] being seduced by the abyss and by the weirdness," Hatem explained.

To draw the audience in, he decided to lean into the disaster angle so often associated with Mothman.

However, the phenomenon is much stranger than a simple warning, and Hatem knew that he needed to illustrate the high strangeness of it all.

"People are trying to figure out what it all means. And it's like, well, what if the universe was trying to warn you, but because it's not human and doesn't share our language, it doesn't know how. So, it's making sounds and it's making weird phone calls, and it's doing this and it's doing that, but only retroactively do you realize, oh, it was trying to warn us about the bridge," he said.

And people responded to that. Although the movie opened to mixed reviews, it has since found its audience within the paranormal community and can be credited with helping popularize the Mothman phenomenon.

"In the movie, people describe feeling really scared about the Mothman. But what's funny is that somehow in real life, the Mothman has […] become this sort of symbol for the outsider, the sexual outsider, the gender outsider; Mothman's out there being weird. Mothman is now a statue with a big muscular ass," Hatem said, only half joking in his assessment of the phenomenon's progression in the popular consciousness.

"And it's been adopted by all of these marginalized communities," he added. "We're just like, great. They're sort of like, yeah, Mothman is us. The thing everyone's afraid of that is hovering in the shadows, but once it comes out of the shadows, it's kind of hot and cute."

But Mothman has spread in many directions since his first popularized appearance in the 1960s, including to Russia, where rumors of a Blackbird of Chernobyl portending the infamous disaster involving that area's nuclear power plant have been circulating online for years. Interestingly, no firsthand source for these stories has ever been found and they seem to date back only so far as *The Mothman Prophecies* movie, in which Hatem wrote a reference to Mothman being seen prior to the Chernobyl meltdown.

"Are you responsible for the Blackbird of Chernobyl? Is that this movie?" Tobias asked.

"I don't know," Hatem answered honestly. "You can understand. It was 1997 when I wrote this. I was not on the internet. No one was on the internet."

While updating events for the movie to take place in the 1990s instead of the 1960s, the Mothman-Chernobyl connection was inserted by Hatem, seemingly from his own imagination. But from there, it appears to have taken on a life of its own, morphing into the Blackbird of Chernobyl.

Chernobyl isn't the only disaster to become retroactively associated with Mothman. Since the popularization of the phenomenon, many such incidents, from a bridge collapse in Minnesota to 9/11, have been connected to the winged cryptid.

"It's like a tulpa," Hatem said, referring to the idea that belief can manifest a phenomenon.

Unlike the Mothman in Point Pleasant, these sightings spread like urban legends after the fact, and if they are driven by belief, then that belief is likely misplaced. Any connection between Mothman and disasters is tenuous at best, and Hatem deliberately focused on that narrative because it made for a better movie, not because of any basis in fact.

When Tobias asked if this was simply a narrative created at the time of the Point Pleasant sightings, Hatem responded, "Yeah, I think that it is, and I think that's certainly the way I used it. I don't know of any supernatural phenomenon where everyone who has seen this particular thing has died tragically, thank God. The thing that pushes it in that direction is that the people who do see something have that feeling of dread and just deep, horrible terror and fear. So, because those sightings are associated with those emotions, it's not hard to go another step and go, 'Well, yeah, boy, if you see that, man, it is bad news, bad moon rising, something bad's going to happen.'"

This is in stark contrast to the silly, or even sexy, way that Mothman is portrayed today in internet memes, although, the screenwriter said, "I'm glad that certain communities have reclaimed it and are having fun with it." He laughed. "I don't know if you're aware, but my friends never tire of sending me the millions of pages of Mothman pornography."

But this lightness isn't entirely a modern invention. Hatem added that humor was something Keel told him he had used deliberately to offset the often dire circumstances involved.

"I think that's kind of nice that it balances out," Hatem said of the humor found throughout Mothman culture. "John Keel told me early on the challenge of writing the book, *The Mothman Prophecies*, was that it was so grim, so dark, and people died, people in the community, people he knew, which is why he tried to write the book with moments of humor and some lightness just to balance it out a little bit. I found that really interesting. It reminded me that, wow, a human being went through this, even secondhand, and then others went through it firsthand when family members died on the bridge or people had experiences that couldn't be explained, that left them with emotional scars that went on for the rest of their lives."

This emphasis on the human element of the phenomenon could contribute to its persistent representation.

"Do you think when it comes to Mothman—specifically the way that Keel wrote *The Mothman Prophecies*, and obviously the way you translated that into the screenplay has a very heavy human element—do you think that that has contributed to the staying power of this monster? Is that what makes it so endlessly fascinating to people?" Tobias asked.

"Well, I hope so, because that's all we have right now. We have no Mothman body, we have no Bigfoot body, we have no crashed discs. At least as far as you and I know, none of this stuff exists in physical reality. All we have are human beings who have had an experience. And that's the interesting part," Hatem responded. "[…] The fact that it becomes a human story is the only thing that's interesting to me."

It is this interest in human experience that has led Hatem down the middle path, becoming neither a credulous true believer nor a cynical debunker. Admittedly, he has never himself had what he would consider a paranormal experience.

"There have been plenty of times I've freaked myself out, but I've never legitimately gone, 'Okay, now I am in the experiencer camp.' Never. And so maybe that makes it easier to walk the middle path, or maybe other people who have had no experiences, it makes it easier for them to say those people are crazy," Hatem said.

However, despite his lack of personal experience, he remains open-minded.

"My opinion of experience is that they have experienced something in general. For instance, in the UFO abduction scenario, I do believe that people have had an experience. I don't know what that experience is though," Hatem continued. "[…] Do I think the people are lying? No. Do I think they're crazy? No. Are they experiencing something? Yes. What are they experiencing? I don't know. Most of them don't know either. There we are. And having not had an experience, I don't feel it's incumbent upon me to define it for them or for myself."

It's that liminal space, between physical and metaphysical, waking and dreaming, fear and wonder, belief and disbelief, that defines this phenomenon. This space has captured the imaginations of millions of people who choose to dwell there, among the monsters, and it is here we must be willing to put down roots if we want to have any chance of understanding Mothman. Where else would we find him, if not his home?

CHAPTER 14
MOTHMAN COUNTRY

Writing a book about Mothman without visiting the city of Point Pleasant would be nothing short of unthinkable. Perceived by many to be the ground zero of the entire phenomenon, this relatively small West Virginia settlement of a little over four thousand people sits on the banks of the Ohio River and the Kanawha River, where the two fast-flowing waterways meet.

On September 19, 2024, the authors of this book, alongside two fellow researchers named Erin and Michael Taylor, set out from their homes and agreed to meet up in Point Pleasant a couple of days before the city threw its annual Mothman Festival.

Driving out from his home in Madison, Wisconsin, Tobias had little difficulty traveling to West Virginia in his car. His three colleagues were in for a much rougher time. They flew out from Sheboygan, Wisconsin, on a short hop to Chicago's bustling O'Hare International Airport. From there, the trio caught a connecting flight into Charleston, West Virginia.

This second leg of the journey was to prove rather more eventful.

Erin worked in the field of mental health therapy. In her spare time, she wrote books about the weird and wonderful folklore of her home state, Colorado, and on other paranormal matters. Her husband, Mike, was a police officer with a sideline in 3D printing curios and sculptures. To commemorate the Point Pleasant trip, he had 3D printed a batch of Mothman statuettes. Each one had the characteristic glowing red eyes; they ranged in size from three to ten inches.

Mike gifted one of the smaller Mothmen to Richard, who accepted it gratefully. The curio would, Richard thought, serve as a pleasant reminder of the fascinating research trip on which the group was about to embark. It looked cute, and Richard had it in mind to be a fun little mascot and keepsake of the research trip.

Tired from the night before, which they had spent investigating the haunting of Sheboygan Asylum, Richard, Mike, and Erin boarded the plane, buckled their seat belts for take-off, then took a moment to pose for pictures with the Mothman statuette. Once the plane reached cruising altitude, they leaned back in their seats, closed their eyes, and dozed.

The first forty-five minutes were uneventful, apart from some very mild turbulence. Then, suddenly, the aircraft's cabin was filled with a loud howling noise. Richard and Mike, who were sitting in rows three and four, blearily woke up and tried to figure out what was going on. The lone flight attendant was sitting in her seat in the galley. Although doing her best not to show it, the sense of alarm she felt was evident on her face. She removed the intercom handset from its cradle on the wall and spoke into it, struggling to be heard over the sound of air escaping from the aircraft.

The plane started to descend. A few moments later, it banked to the left, going into a shallow turn. The captain spoke over the intercom. Sitting farther back in the passenger cabin, Erin was able to hear him talking, but from their ringside seats, Richard and Mike couldn't make out a thing over the sound of rushing air.

When the jet reached a lower altitude and leveled off, the pilot tried again. He explained that the door in the galley seemed to be having a mechanical problem and that for everybody's safety, he had made the decision to turn the flight around and return to Chicago O'Hare, rather than risk pressing on for Charleston. Nobody had any issues with that. Pilots are masters of risk-benefit analysis, and if he was sufficiently concerned with the air leak to turn back, nobody was going to argue. After all, it's better to be inconvenienced than to be dead.

The plane landed safely and taxied to its starting gate. The passengers deplaned. In a commendable show of efficiency, the airline had a replacement aircraft staffed by the same crew on hand in just under an hour. Richard, Mike, and Erin made it safely to Charleston with only a minimal delay. Everybody was relieved that the in-flight emergency hadn't been worse.

Taking the little Mothman statuette out of his pocket, Richard found himself pondering the cryptid's status as a perceived harbinger of doom and tragedy. Surely there was nothing to it … right?

It is no insult to say that Point Pleasant is one of those cities that has a single card to play, and boy, did it play that card well.

The card, of course, was Mothman.

That's very apparent from the moment you first enter town. Mothman was everywhere … literally *everywhere.* The creature appeared on the side of buildings; in shop windows and on storefront signage; as we drive into town, the crew spotted more than one This House Is Protected by Mothman signs posted on the outside of private residences. These were meant in humor … most likely.

It was late on a Thursday evening when the group found a parking spot and made their way into the center of the city. The sun was still up, but most of the businesses that catered to tourists were closed. All around them, preparations were being made for the annual Mothman Festival, which was expected to draw thousands of visitors to Point Pleasant over the following three days. There wasn't a hotel room to be had for miles around. The team knew this because Richard had called them all and asked.

Tents and shade structures were being placed at street intersections. Food trucks were parked and set up in anticipation of feeding a horde of hungry people over the course of the weekend.

The group of four knew that there would be plenty of time to explore the city, but first things first. They wanted to do what every new tourist seemed to do when they arrived in Point Pleasant—pay a visit to Mothman. Or the next best thing, his statue.

The twelve-foot-tall piece of art had lived in the city center since 2003, situated next to the famous Mothman Museum. Sculpted by artist Bob Roach, this incarnation of Mothman looked as though he was no stranger to the gym. Abs of steel and sculpted pecs stood out as the creature leaned forward, its menacing, claw-tipped arms spread wide. Two huge ruby-red eyes lent the statue's face an ethereal look. When the light caught them just right, they glowed.

Bob Roach died in 2015, but his legacy lived on in the form of his most famous creation. As they crossed the street and headed toward the statue, it struck the group just how impressive a piece of art this was. Each took their turns to stand alongside it, striking poses for the camera that ranged from serious and somber to frivolously silly. A kindly passerby volunteered to take a group shot.

The less said about the now popular tradition of inserting a quarter between Mothman's buttocks, the better. Suffice it to say that this has now become a thing, to such an extent that at least one T-shirt was on sale that advocated that very practice.

Once the pictures had all been taken, the team retired to its accommodation for the night. They were back in Point Pleasant at 10:00 a.m. the next morning. That was the scheduled opening time for the Mothman Museum, and they wanted to check it out before the big crowds arrived the following day.

It was, all four of them agreed, easily worth the five-dollar admission charge. The museum made an earnest effort to tell the Mothman story in a colorful and engaging way. For the most part, it succeeded. Newspaper clippings, many of them blown up for easier reading, adorned the walls, documenting the 1966 to 1967 sightings and related events. The collapse of the Silver Bridge was covered extensively, and so was Mothman's place in popular culture. There was also an extensive collection of movie memorabilia from *The Mothman Prophecies,* including costumes and props from the production, many of which were recognizable from the screen.

Museum visitors could pose with various mannequins and statues of the city's infamous cryptid. Mothman dangled from the ceiling, looming over customers as they entered the museum via the gift shop. There was a model of the north power plant, the long-abandoned building where Mothman was initially sighted. Some have speculated that the creature was actually nesting there during part of the initial sightings flap.

Richard gleefully fessed up to posing for photos with as many different incarnations of Mothman as he could. After half an hour, the group had seen all they wanted to see and made their way out to the shop. The shelves were bursting with every piece of merchandise that could conceivably bear the likeness of Mothman: T-shirts, posters, mugs, models, pins, fridge magnets... the list went on. Even two days before the festival proper was set to begin, the inventory was practically flying off the shelves.

Having crossed the world's only Mothman museum off their bucket list, the team's next stop promised to be considerably more somber. It would be easy to miss, unless the visitor was informed and paying close attention to their surroundings. Painted on a section of the city's floodwall overlooking the Ohio River was a colorful mural, which was painted by artist Jesse Corlis in 2018. Standing in front of it gave one the impression of looking directly across the old Silver Bridge. In the background, heading away from the Point Pleasant side of the bridge, was an automobile dating back to the 1920s. In the foreground, driving toward the city, was a pickup truck circa the late 1960s. These two vehicles represented the beginning and the ending of the forty-year lifespan of the Silver Bridge.

Tobias, Erin, Mike, and Richard all stood before the mural, just taking it in and appreciating it for the work of art that it was. Only later would they learn that the forty-six birds that flew across the painted sky represented those who died in the tragic collapse.

It was impossible not to be moved. On the opposite side of the wall, looking out toward the still-standing railroad bridge, was a formal marker indicating that this was the site of the Silver Bridge. Some nameless kind soul had left a pebble atop the memorial, with five painted birds and the phrase "you are not alone."

The group approached slowly, with reverence. At their feet, the names of those forty-six lost lives were tastefully engraved on memorial bricks.

All four researchers fell into a respectful silence. At least one of them began to tear up. They spent a few moments quietly contemplating the import of that place, each lost in their own thoughts and simply paying their respects.

What happened there on the evening of December 15, 1967, was important. It was important to the history and character of Point Pleasant, because those events still scarred the town in the present day. It was important because forty-six lives were lost, nine people were injured, and a significant amount of emotional trauma was inflicted upon the survivors, the bereaved they left behind, and the rescuers who tried to save the victims of the tragedy. As such, the tragedy bears closer examination.

Completed in 1928, the Silver Bridge was built with the express purpose of connecting townships in Ohio and West Virginia. It linked Kanauga and Gallipolis on the Ohio side of the river with Point Pleasant on the West Virginia bank. Why call it the Silver Bridge? Because of the aluminum paint, a new concoction at the time, which made the bridge gleam in a unique way. Costing just under a million dollars to build, the people who lived on both sides of the bridge were understandably proud of it. In terms of its design, this was the first span of its kind in the United States. That design would later come to be its undoing.

That undoing came shortly before 5:00 p.m. on the evening of December 15, 1967. The Silver Bridge was heavily laden with traffic. Not only were commuters coming home from work, but with Christmas just ten days away, holiday shopping was in full swing.

The bridge was supported by a series of eyebars, which were composed of a series of support chains kept under tension. Only months later would investigators learn that a single weakened eyebar located on the section of bridge closest to Ohio had broken, after having been subjected to years of corrosion and stress. Like dominos falling one by one, this led to a catastrophic failure, which brought down the entire bridge. Vehicles plunged into the frigid depths of the Ohio River or were smashed into the ground as they patiently waited to cross the span.

Once the collapse was complete, all that remained standing were the vertical supports. Many of those unfortunate people who went into the water drowned. Some of their bodies would not be recovered for weeks. Other victims located closer to dry land died of major blunt force traumatic injuries.

There were many heroes that night. First responders raced to the scene but were soon overwhelmed by the sheer scale of the carnage. Victims needed to be pulled from the water. Some clung to floating objects. Others were trapped in their vehicles and had to be extricated. Civilians of all ages pitched in to help.

Once the mass rescue operation was underway, the local hospitals were overwhelmed with the injured and the dead. The military was brought in to assist. Combat engineers arrived with specialist training and resources. Temporary morgue facilities were set up, and the federal government stepped in to assist with the rescue and reconstruction efforts.

The people of Point Pleasant would never forget the awful events of that night, but life had to go on. Uncle Sam fast-tracked the design and construction of a new bridge, located farther along the river. With a respectful nod to the tragedy, the Silver Memorial Bridge was opened to traffic on December 15, 1969—two years to the day after the disaster. By common consensus, the design used for the replacement bridge was a more traditional one. There were no eyebars to be seen anywhere.

This is the bridge that Tobias, Erin, Mike, and Richard drove across on their way into Point Pleasant for the first time.

But, what of Mothman?

In *The Mothman Prophecies,* John Keel wrote about his companion, Mary Hyre, suffering a nightmare that appeared, with hindsight, to be more of a premonition. She had dreamed of people drowning in the river, and of abandoned Christmas gifts floating downstream. According to Keel, this took place in November 1967, a month before the bridge collapsed.

Keel devoted the final stretch of his book to the disaster. He did not explicitly link Mothman with the Silver Bridge collapse, stating, "There's no answer to that."[70] However, he went on to say that he was warned that something terrible was about to happen, and that if he had only put those warnings together in time, then many lives might have been saved.

---

70. Keel, *The Mothman Prophecies, 294.*

Without the gut punch that the Silver Bridge tragedy delivered, the book's ending would have had significantly less of an emotional impact. It delivered the major action set piece in the 2002 movie of the same name. The scene evoked a sense of creeping dread and then outright horror as the bridge slowly came apart.

In reality, any connection between the Mothman sightings and the disaster was speculative. As they walked around Point Pleasant over the next few days, the team would see a lot of artwork depicting Mothman and the bridge together. They had become inextricably joined in popular culture, even though there was no evidence to suggest that either Mothman or the UFO phenomena reported throughout the locality had anything to do with the bridge going down. The likeliest explanation was also the simplest: a stressed component forming a single point of failure, and a flawed engineering design, came together after four decades of wear and tear to cause a tragedy.

After paying their respects to those who had died on the Silver Bridge, the team took some time out to wander the streets of Point Pleasant and get a feel for the place. Setup for the next day's festival was in full swing. Tobias and Richard were particularly interested in speaking with local people and gaining their perspective on things. They lucked out after ducking into a store, where Richard bought a small Mothman pin from a nice lady named Jan.

Jan had lived in the vicinity of Point Pleasant for her entire seventy-two years of life. She was very chipper and fully willing to share her opinions on Mothman and the festival.

"We had fourteen thousand people last year [in 2023]," Jan gushed, ringing up Richard's purchase and accepting a five-dollar bill. "Mothman is great for business. It brings in a lot of people and it's a heck of a lot of fun."

Richard asked whether her upbeat view was representative of that held by her friends and neighbors. She told him that most of the people of Point Pleasant chose to stay home that weekend, but that many still appreciated the benefits such an influx of tourism brought.

"Without Mothman, Point Pleasant would be nothing," she stated flatly. "Absolutely nothing. Back in the days before the bridge collapse, we had a lot of stores here in town and over the river in Gallipolis. Clothing, shoes, jewelry. Now they're practically all gone. Without Mothman, nobody would come here."

To prove her point, she swept an arm around the store. There were Mothman comic books, art prints, statuettes, action figures—you name it, they had it, all of it emblazoned with the Ohio River Valley's most famous red-eyed cryptid.

"You've lived here since 1951," Tobias observed. "That's years before the Mothman sightings began. What was it like around here in 1966 to 1967 when Mothman fever was at its peak?"

Jan was a sophomore in high school at the time. She told the two researchers that what would later come to be called Mothman, most likely because of a newspaper reporter who needed a catchy name for the creature, was thought at the time to be a giant blackbird.

"In fact, my sister saw the thing, red eyes and all, one night when she was parked by the TNT area. Of course, we don't know what she might have been smoking or drinking…" An amused twinkle gleamed in Jan's eye as she alluded to the TNT area's reputation as a place for young people to hang out at night and enjoy some quality private time together. "There were many sightings of that big blackbird around that time."

She attributed the arrival of the great bird to the curse supposedly issued by Chief Cornstalk. This was one of the many folkloric legends that could be found in this part of the country. In the mid-1770s, Chief Cornstalk was a leader of the Indigenous peoples who opposed the colonists as they spread westward, increasingly encroaching upon Native territories. His Shawnee name was *Keigh-tugh-qua*, which approximated to "cornstalk" in the English language.

It was inevitable that the two factions would clash, and on October 10, 1774, Indigenous warriors of the Confederated Tribes faced off against the militia in the Battle of Point Pleasant. The fighting was bloody and desperate, with heavy losses on both sides. The forces under Cornstalk's command retreated from the field in what was ostensibly a defeat.

Although deeply unhappy about it, Cornstalk ultimately signed a treaty with the whites. During the Revolutionary War, he and some of his people were taken as hostages by the colonials in an attempt to prevent the Shawnee from siding with the British. They were treated well, but things took a tragic turn when two colonial soldiers were set upon by tribal soldiers while on a hunting foray. One was shot dead.

Outraged comrades of the dead man murdered the completely innocent Cornstalk and his companions in cold blood. Before he died of multiple gunshot wounds, the tribal leader supposedly placed a curse on the land around him and those who would come to inhabit it. Shortly afterward, that land would be the site of a new settlement—a place named Point Pleasant.

Although they were moved more than once, in 1954, Chief Cornstalk's remains were interred in a monument located in the Tu-Endei-Wei State Park in Point Pleasant, close to the site of the battle. They remain there to this day. Although he is long dead, belief in Chief Cornstalk's alleged curse remains alive and well in the region today. The curse has been blamed for almost every calamity or significant misfortune in the region for decades. Plane crashes, vehicular collisions, and industrial and mining tragedies have all been attributed to Chief Cornstalk's thirst for vengeance, as though the dead man were somehow reaching out from beyond the grave to take retribution centuries after his murder.

Needless to say, the Silver Bridge disaster topped the list. So too did the arrival of Mothman, or the big blackbird, as Jan pointed out to her customers.

"I believe that it was sent by Chief Cornstalk to warn us that something bad was going to happen," she explained.

"But if you're going to curse someone, why would you then send a warning to them before disaster struck?" Richard asked. It didn't seem logical to him that a long-dead tribal chief would seek vengeance by leveling a curse, only to then send a harbinger to alert the cursed. Why telegraph the punch before it landed?

"I don't know," Jan admitted, pondering the question for a moment. "Maybe he kind of ... felt bad?"

"Was the curse something that people were talking about in 1967?" Tobias wanted to know. The answer from the lady who was actually there at the time was an unequivocal no. "So, that all came about after John Keel published his book ... "

"Exactly," nodded Jan. "Everybody just thought it was this strange, big blackbird. I thought it was just a crane that had somehow gotten lost from its flock."

Her memories were borne out by newspaper articles from that time, many of which described the creature as either a "thing" or a "bird." A November 20, 1966, article in the *Athens Messenger* printed a picture of a sandhill crane alongside the opinion of Dr. Robert Smith, an associate professor of biology at West Virginia University ("Is Mysterious Creature Balloon or Crane?"). This was based on the premise that sandhill cranes can reach the size of an adult human.

However, perhaps the best hypothesis in that particular article came from an unidentified man—probably unidentified on purpose in an effort to ensure his safety—who told the reporter that the description was an exact fit for his mother-in-law, "especially the red eyes."

Well played, sir.

However, those who actually set eyes on the creature were adamant that whatever they saw was definitely no bird.

Other contemporary accounts referred to it as the Mason County Monster or simply a Red-Eyed Whatever. These names would eventually become supplanted by Mothman, a moniker that reigns supreme to this day.

"I certainly didn't know about the curse back in the 1960s," continued Jan. Tobias now had reliable confirmation that nobody in Point Pleasant or the surrounding area was connecting the Silver Bridge collapse with the arrival of Mothman. The narrative of Mothman being a harbinger of disaster would not arise until much later, following the publication of John Keel's book.

"Of course, some people do say that West Virginia is a kind of portal to the unknown," she mused, placing the pin in a small bag and handing it to Richard, along with a receipt. "Once the Silver Bridge fell, the bird was gone. There were no more sightings... and we don't ever want to see it again!"

By Friday afternoon, the small stream of visitors trickling into Point Pleasant in advance of the Mothman Festival had begun to significantly increase. The team chatted briefly over lunch in the town's sole Mexican restaurant. Tobias tucked into his first (and potentially last ever) Mothman burrito, which came complete with baked wings and sliced tomatoes for eyes. He and his three companions all agreed that they wanted to hit one of the most iconic locations associated with the Point Pleasant sightings before it became too crowded with sightseers: the infamous TNT area.

Although not readily signposted, the TNT area wasn't all that difficult to find. It formed part of a wildlife preserve located a few miles outside of town, and a small part of it was still publicly accessible. Mike drove. The team knew that they were close when they saw the word *Mothman* spray-painted on the road, along with an arrow pointing off toward a long track that was bordered with woodland on one side and swampy marshland on the other.

Some people have suggested that the long-abandoned concrete igloos, once used to store TNT, might have been Mothman's lair. There's no evidence to support that contention; indeed, the events of the 1966–67 sightings imply that the now demolished brick power plant, which was once located in this area, might have fulfilled that role, as the visit to the Mothman museum had reminded the team. Mothman was sighted on and around that particular building, which was reduced to rubble back in the 1990s. (Of course, it's entirely possible that Mothman never had a lair at all.)

Each igloo was basically a blast-resistant bunker, as befitted their original purpose. Ducking inside the first one, the researchers found a morass of graffiti adorning the inner walls of the dome, much of it either obscene, Mothman-related, or, in some instances, both. It was cooler outside the glare of the hot November sun. Every time somebody spoke, the echo reverberated from the walls, essentially deafening everyone inside the silo for the next five seconds before slowly dying away.

Stepping back out into the sunlight, the group walked farther along the track. There was a bunker every few hundred feet or so. Each one looked pretty much the same inside. Finally, they came to one that was open to the sky. The uppermost part of the dome had been completely sheared off. Tobias and Richard believed this was the bunker that exploded back in 2010. Although public details are scarce, what was openly known is that the bunker contained several thousand pounds' worth of volatile materials, which predictably blew up, destroying the top of the igloo in the blast. Federal agencies descended on Point Pleasant in hordes, allegedly to determine whether this was an act of domestic terrorism or not. There was also some local speculation about it having been a potential meth lab.

Apparently, it wasn't; but the TNT area was shut down and cordoned off for weeks. According to a report by the *Associated Press*, the culprit actually turned out to be improperly stored gunpowder.[71]

The quartet was all thinking the same thing. As paranormal investigators, they wanted to attempt communication with any discarnate intelligences that might be lurking around in the area. It was a little too noisy at that time, but Richard had a feeling that things would probably be a lot different after dark. Once they had finished checking out the final igloo, the team reached a T-junction at the distal end of the track. There was nothing else to see. Although there were numerous other bunkers—supposedly around a hundred in total—all of them were located on private land, and nobody was willing to trespass. For one thing, it would have been wrong. For another, it could also have been a great way to get shot, as they'd been hearing sporadic gunfire all afternoon. Absolutely nobody wanted to be the target of an overly zealous hunter.

Retracing their steps along the track, Richard paused when he ran into two friendly men who asked where

71. "Explosion Reported at Munitions Storage Site."

they could find the bunkers. He recognized their accents instantly. They were fellow Brits and were in town for the Mothman Festival. This made Richard curious. It was a long way to travel across the Atlantic, all for a two-day festival. Their names were Miles and John. The two visitors quickly fell into a conversation with Tobias and Richard.

Miles was a hardcore Mothman enthusiast, and John was his dad. Miles was on a pilgrimage, the culmination of a lifelong obsession with the red-eyed cryptid. John wasn't about to leave him to make the journey alone.

"What's the fascination with Mothman?" Richard asked.

"The sheer mystery of it all, and the sense of the unknown," Miles replied without hesitation. "Is it a living being, or something multidimensional? Nobody knows!"

They were determined to visit the site of the Silver Bridge collapse and then to see everything possible in and around Point Pleasant itself. Their goal was to "just take it all in," to gain a new appreciation of where the town fit into the broader Mothman mythos.

"Is there a lot of awareness about Mothman in the United Kingdom?" Richard wanted to know.

Perhaps surprisingly, they revealed that there wasn't. Richard had thought that the global nature of the phenomenon, coupled with Mothman's place in popular culture, would have made him instantly recognizable, yet it transpired that not many Brits have either read Keel's book or seen the movie based upon it. Miles recognized Tobias from his Mothman-related appearances on *Unsolved Mysteries* and *Small Town Monsters.* His distinctive cowboy hat certainly helped.

Miles was clearly well versed in Mothman lore, because he and Tobias quickly fell deep into a discussion about the varying reports about the creature's facial features.

"The specific details should differ depending on the person who sees Mothman, if we're viewing this phenomenon through a Keelian, Ultraterrestrial lens," Tobias pointed out. If he was right, this might have explained why some experiencers describe Mothman as having a feathery face, whereas others recall it as being black or even completely devoid of features. Every different observer would be seeing a different facet of the phenomenon.

"If you think about the original eyewitnesses in this area," Miles mused, "they see owls and other birds all the time. Their brain could have gotten so used to seeing that type of wildlife that, if it's perceiving something as radically different as the Mothman phenomenon and trying to make sense of it, those perceptions might assume a form close to their everyday understanding."

In other words, if you lived in a rural area, and birds such as owls and cranes were part of your everyday worldview, then your brain might reach for those familiar characteristics when trying to interpret something that is completely and utterly alien to normal reality. It was an intriguing concept, and one that might also account for sightings of Goatmen, Dogmen, and other strange beasties or hybrid creatures.

"Where we live in the UK, we don't see as much wildlife as the people of West Virginia do," added Miles. "So, an experiencer who lived in our area might see a fuzzy black cloud with red eyes, just because it lacks that nature-based frame of reference."

Tobias and Richard had never found the "it was just a big blackbird, bat, or sand crane" hypothesis an easy one to accept—at least, not when they come from people who live in the area. Locals see those birds regularly, and in their view, they would be highly unlikely to misinterpret them for being something as bizarre as Mothman.

"Most misidentifications of flying humanoids or creatures—and there are a fair number of totally honest, sincere mistakes—come from people who are not totally familiar with their environment." Tobias was speaking from many years' experience interviewing such eyewitnesses and taking their statements. He recalled the Chicago Mothman sightings and noted that there is a significant number of migratory bird traffic moving through those areas. A visitor to the region who was unfamiliar with such animals might easily misidentify them. "Somebody who's been a city dweller their entire life travels out of state and suddenly sees a great blue heron for the first time, may not know what on Earth it is. Sometimes they think they're seeing a monster. Rural people, on the other hand, tend to know better."

Once again, the investigators were reminded of the fact that Point Pleasant was actually only a relatively small part of the Mothman story. The 1960s wave of sightings took place throughout the Ohio River Valley, with the big blackbird / Mothman being seen by close to a hundred eyewitnesses. These are people who know what the birds that were native to the area looked like, because they encountered most of them daily.

"We spoke to a resident in Gallipolis yesterday," John nodded. "He dismissed the whole Mothman phenomenon out of hand... blamed it on giant owls that live in the area, owls with six-foot wingspans. I asked him about the red eyes, and he said that if you shine a flashlight on them at night, they'll glow red."

"Talk about classist nonsense!" Tobias snorted. "Certain elitist segments of society refuse to take rural people seriously and won't believe in the authenticity of their experiences on any level. They assume, 'Oh, they're rural so they must be stupid!' The skeptical explanation becomes that what the experiencer *actually saw* was an animal with which they are already perfectly familiar... something they see all the time."

A case in point would be that of the 1955 Kelly-Hopkinsville Goblins, in which a gaggle of alien-looking humanoid creatures is said to have laid siege to a farm in rural Kentucky. The family who lived in that farm predictably opened fire on the strange beings with everything in their not inconsiderable arsenal, putting out munitions like they were going out of business. The "goblins" were described as having long arms with talons at the end; thin, lipless mouths (not beaks); glowing yellow eyes; and large, upswept ears.

Skeptics have claimed that what the multiple experiencers actually fought against were none other than a bunch of owls—great horned owls, to be precise. If that was the case, the family in question must have been extraordinarily bad shots, because not one dead owl carcass was found at the scene when police arrived after the firefight to investigate. Never mind the fact that people living in rural Kentucky at that time almost certainly saw great horned owls on a regular basis and never mind that there was a reported UFO sighting immediately prior to the encounter.

Rural people must have been drunk, stupid, or both. Their experience could not possibly have been a brush with the bizarre. Such things don't happen, because such things can't happen. QED. Case closed ... or so say some of the skeptics.

Miles, for his part, remained undeterred. His goal in coming to Point Pleasant was a simple one, and to him, it was also exceedingly meaningful. "I would like to have a sense of this space," he explained. "To get a feel for what the area is like ... the people, and the town. Take it all in. To gain a better understanding of what happened here all those years ago."

It was a commendable goal, and as the four men shook hands, Tobias and Richard wished the two pilgrims well. They were all engaged on a similar quest, after all, and everybody wished for the others' success.

After parting ways, they trudged back to the car and made the drive back into Point Pleasant for dinner. The TNT area and its mysteries would all keep until sunset.

They had no idea of the strangeness that awaited them.

CHAPTER 15
TWILIGHT TIME IN THE TNT AREA

After having eaten, the team of four returned a few hours later, at dusk. Apart from the sound of nocturnal critters going about their business in the trees and the marshes, the TNT area was busier than it had been earlier in the afternoon. More sightseers had arrived, and the sounds of raucous laughter could be heard coming from somewhere within the vicinity of the igloos.

Forestalling any misplaced irritation on the part of his companions (after all, those visitors had every right to be there), Tobias noted that during the Mothman flap of 1966, right after the initial Scarberry-Mallette encounter, that particular part of the TNT area was also filled with people.

Sightseers and thrill seekers had flocked there at night, many of them staying overnight in the hopes of catching a glimpse of Mothman himself—or of the strange lights in the sky that accompanied the creature.

"They came out in droves," Tobias announced, "invading private property, doing anything they could to catch a glimpse of the unexplained."

After a while, things began to quiet down. The shadows lengthened. Breaking out a state-of-the-art thermal imaging camera, Tobias led the way into the deepening dusk, sweeping its lens methodically from left to right and back again as the team walked along the track.

Walking past a gap in the trees, Richard's heart leaped into his mouth when he saw what, at first glance, appeared to be a pair of glowing red eyes caught in the beam of Mike's flashlight. On closer inspection, it turned out that some wag with one heck of a sense of humor had taped a pair of circular reflectors to a tree, presumably with the intent of giving an unsuspecting visitor the shock of their life.

Whoever was responsible, if you're reading this ... well played.

There was no point in going into any of the enclosed bunkers. The echoes would only contaminate any audio recordings the researchers made. The open igloo was another story, however, and they dropped their packs on the ground just inside the entrance. There was nobody in that part of the TNT area, so Richard, Tobias, Erin, and Mike staked their claim for the next few hours and settled down to the business of trying to make contact with ... something.

The remnants of a rusty old box mattress filled the center of the bunker. Somebody had apparently set it on fire, as the only bits left intact were parts of the wooden frame and several springs. It made for a convenient, if somewhat rickety, place to set down a few pieces of equipment.

Richard put down a Dead Bell, a device that senses electromagnetic energy in the environment and makes a bell chime in response. Some paranormal investigators believe that spirits of the dead manipulate the bell to deliver answers to yes-no questions. As far as Richard was concerned, the jury was still out on that claim, but his logic was that if it *might* be able to communicate with the spirits of the dead, then who was to say it couldn't also be manipulated by some aspect of the Mothman phenomenon?

That's not to say the team believed Mothman would literally be out there—as exciting as that would have been. No one thought they were going to run into the red-eyed winged cryptid in person ... but perhaps some remnant of the strange events from 1966 to 1967, some lingering echo, might still be around and capable of making its presence known. They could but hope.

Tobias scanned the woods surrounding the open-air igloo with his thermal imaging camera, looking for critters, humans, or anything else out there. He was also keeping an eye out for light anomalies.

"According to Keel, Ultraterrestrials look nothing like we actually observe them," he said, sweeping the camera back and forth. "What they might *actually* look like—their true forms, if you will—could be those anomalous balls of light."

Reports of such balls of light go back far into the annals of paranormal research, and cross boundaries between different types of phenomena. Luis Elizondo, the former director of the US government's AATIP (Advanced Aerospace Threat Identification Program) noted in his autobiography *Imminent: Inside the Pentagon's Hunt for UFOs* that mysterious balls of light—Elizondo refers to them as orbs—began appearing inside his home and those of his colleagues at AATIP.[72] Other members of his family witnessed them too. The phenomena even bled over into the homes of Elizondo's neighbors. Ominously, he describes being near these light anomalies as having had "severe biological effects resulting in life-threatening medical issues" on individuals Elizondo knew and worked with. The precise nature of those ailments remains classified, but if these are indeed connected to John Keel's Ultraterrestrial hypothesis, then it would behoove anybody coming into contact with them to give the orbs a wide berth.

"You would not believe how many reports of balls of anomalous light I get," Tobias replied when Richard pointed

---

72. Elizondo, *Imminent*, 68.

this out to him. "They are associated with every sort of phenomena imaginable, ranging from ghosts and hauntings to cryptid encounters and UFO sightings—including alien abductions and shadow people."

Erin, Mike, and Richard were all resolved to keep an eye out for any strange lights in the vicinity tonight. Erin had set out motion detectors and electromagnetic sensors, while Mike placed a series of vibration-sensing devices around the igloo.

It wasn't long before they realized the Dead Bell wasn't going to be of much use in that environment. The device was pinging at random intervals, completely unrelated to the sequence of yes-no questions they were asking. Even with the sensitivity dialed down, it was clear that all the team was getting out of the device was a bunch of false positives. The Dead Bell had delivered intriguing results on prior investigations, but those had always been indoors in a relatively controlled environment. This was the first time Richard had employed it out in the open air, and it seemed he had brought along the wrong tool for the job. Reluctantly, he switched it off after one final peal echoed around the concrete walls of the igloo and then died out.

Erin and Mike got set up for the next experiment, a technique paranormal investigators refer to as the Estes Method. This involved one person—in this case, Erin—being blindfolded and wearing a set of professional-grade headphones that are connected to a spirit box. The spirit box is essentially a radio scanner, hopping from frequency to frequency on the AM or FM band. Proponents believe that spirits of the dead speak to the living in the spaces between frequencies. Whether this is true or not remains a matter of contention within the paranormal community, but the technique has many advocates, and the team was fascinated to see whether they could establish contact with something intelligent there in the blast-damaged silo.

There were no lights whatsoever in the silo. Looking up, Tobias and Richard spotted a bright pinpoint of white light arcing across the sky among the constellations. It was clearly a satellite, but it put Richard in mind of what those hopeful UFO experiencers must have felt back in the 1960s when they came out in search of extraterrestrial visitors. A shiver ran down his spine, caused by a sense of connection with people of like mind who walked here more than fifty years ago.

Erin set the frequency sweep rate to her liking and put on the blindfold and headphones.

"Based off Keel's Ultraterrestrial hypothesis, the phenomenon Mothman represents may be capable of communicating via the medium of consciousness," mused Tobias. "So, techniques like this, which are meant to connect with discarnate intelligences, could potentially be very useful in our research."

Erin squatted down on her haunches next to the doorway, entering a state of sensory deprivation. Ever the protective husband, Mike leaned against the opposite wall, keeping a careful eye on both her and the darkness beyond the door.

Everybody in the silo fell silent. Tobias continued to scan the darkness with his infrared and thermal imaging camera. Mike continued to stand watch. Richard divided his attention between watching the satellites move against the starry backdrop above and monitoring the vibration sensors.

For the first few minutes, nothing at all came through the spirit box: no commercial stations or private radio traffic. Then came snatches of words. Fragments. Nothing meaningful. After a while, Erin changed frequency bands, switching from FM to AM.

Richard began asking questions, hoping for a response.

"Can you see the same night sky that we can? All those stars?"

The answer seemed like a meaningless number: twenty-two. Tobias shrugged. The timing made it seem like a response, but that wasn't for sure.

"Something very strange happened here back in 1966," Richard continued. "Do you know anything about that?"

"I heard the name Mike," Erin said, repeating words she'd just heard spoken to her via the headphones. Mike—*our* Mike—didn't say a word.

"What is it about this place, or the situation, that you want us to understand?" Tobias asked, his eyes still glued to the camera screen.

Erin was hearing the voices of both a female and a male. Their response to Tobias's question was, "We can laugh and pretend."

Richard followed up with some specific questions pertaining to Mothman, but the two voices now deteriorated into little more than indistinct murmuring. The team let it run on for a while longer, giving the voices an opportunity to stabilize, to come back. They never did.

Finally, Erin slipped off the headphones and blindfold. It was a valiant effort, but the team hadn't been able to make sustained contact with anything intelligent. The hour was growing late, and there was still more research to be performed. None of the sensors had been tripped. Finally, the four researchers packed everything up and started making their way back to the car.

As they reached the track, something strange happened. An odd white light was flitting around the TNT area, flashing through the trees and across the swamps. At first, the team thought it might have been car headlights, or possibly a fellow Mothman enthusiast shining a flashlight around. Except there was nobody out there. Tobias confirmed it with his thermal and infrared imager, which had an effective range of close to a thousand feet. There was nobody hiding inside the tree line, or in any of the igloos that were passed along the way.

As abruptly as it arrived, the mysterious light was gone again. Richard, Tobias, Erin, and Mike were all very aware of the conversation they'd just had an hour ago, concerning light anomalies and their potential link with Keel's Ultraterrestrial hypothesis. Could they have just encountered this for themselves, or was there a more mundane explanation?

If there was, they were unable to come up with it, and it puzzled all four of them during the drive back to Point Pleasant.

CHAPTER 16
MOTHMAN PARTY

If you ever decide to visit the Mothman Festival—and the team absolutely recommends doing so—then their advice is to get there *early*. Thousands of visitors arrive in the town over the course of the weekend, with the majority of them turning up on Saturday morning. Enterprising locals, businesses, and churches rented out parking space in their yards for twenty dollars a vehicle.

Mike parked on the periphery of town, donating parking money to a church fund in exchange, and the group strolled into Point Pleasant. It was late September, and unseasonably hot. Although the thermometer read in the nineties, the direct sun beating down was brutal, especially when coupled with the humidity. Pretty soon, all four were sweating from every single pore in their bodies.

A huge inflatable Mr. Stay-Puft, the Marshmallow Man from the movie *Ghostbusters*, towered above the crowds outside the town theater. Main Street was a packed, jostling mass of humanity. Visitors were teeming around every vendor stall, purchasing all manner of Mothman-themed merchandise. It was good to see other cryptids represented too. Richard spotted a T-shirt bearing the image of the pterodactyl-like Van Meter Visitor, bringing back memories of his recent Midwest adventure with Tobias.

Bigfoot was also popular, as were some lesser-known mythical creatures such as Wisconsin's Hodag. A small crowd of children were enraptured by an animatronic Hodag set up inside a packing case. It rattled, roared, and then emitted an impressive amount of steam. It was heartwarming to see the sheer wonder on the faces of those kids, a timely reminder of that spark of curiosity and fascination that gives strange beasties such as the Hodag and Mothman their timeless and ageless appeal. Based on the sheer number of octogenarians on the packed streets of Point Pleasant that weekend, it was clear that whether you were eight or eighty, Mothman will always have the capacity to enthrall and intrigue.

The people-watching was fun. There was a lot of joy to be had in seeing thousands of human beings coming together out of a shared love for something as esoteric as Mothman. At lunchtime, the foursome boarded a bus to experience one of the town's Mothman-centric tours. The air-conditioning came as a blissful relief from the heat and humidity.

"Y'all ready to go hunt Mothman?" the guide asked rhetorically as the bus pulled away. She was answered with a resounding cheer from the customers. Over the next hour, they were driven out to the TNT area and given an overview of the Mothman phenomenon as it related to the city of Point Pleasant. Not much was said about the bigger picture, which was understandable given the time constraints and relatively casual nature of the tour. The information presented was mostly consistent with what's widely known and was sprinkled with some tidbits of local knowledge and a few vignettes to add a little flavor.

On the way back from the TNT area, the bus passed along the stretch of road on which the Scarberry-Mallette encounter had taken place. It looked ordinary enough, but after having read about it for so many years, Tobias and Richard were both excited to finally see it for themselves.

It was early afternoon when the bus deposited its passengers back in Point Pleasant, just in time for the researchers to meet with their first interviewee of the day.

In all their years spent researching ostensibly paranormal phenomena, Tobias and Richard have interviewed many individuals. Those stories have ranged from the slightly strange to the downright bizarre.

Few of those shared experiences can compare with that of Mark May. They caught up with Mark at his vendor stand, located high above the riverbank just a stone's throw away from the site of the Silver Bridge collapse.

Ever since he was a boy, the forty-five-year-old Mark had been obsessed with that bridge. December 15 has always been marked on his calendar as Silver Bridge Day for as long as he can remember. As a young man, Mark visited a psychic with the intent of asking about his past lives. The psychic fixed him with a somber glare and told him, "You know perfectly well who you were, Mark…you were a victim of the Silver Bridge disaster."

That didn't come as a great revelation, because as wild an explanation as it may sound, it also felt intuitively right to Mark. He does have a strong idea as to the identity of that person, but out of respect for their next of kin, it will not be revealed here.

The bridge collapse was a common topic of conversation among members of Mark's family when he was growing up. As an adult, he took to visiting Point Pleasant and walking along the riverbank, inextricably drawn there. He even owns a piece of the bridge itself, which was retrieved by his grandfather and now lives in Mark's rock garden.

"I don't feel sad or melancholy when I come out here," he explained, sweeping out an arm to indicate the river below. "I just feel a deep sense of connection with Point Pleasant and the things that happened here."

One of the people Mark encountered in Point Pleasant was a former World War II OSS (Office of Strategic Services) veteran, who matter-of-factly told him of his own

Mothman sighting. While working at the National Guard Armory, located northwest of Ohio River Road, he once saw an enormous blackbird, comparable in size to a light aircraft, with "flapping wings and manlike legs" dangling beneath it. The creature flew in over the treetops, approaching from the direction of the Ohio River. Looking at the site on a map, it became apparent just how close the armory is to the TNT area.

Crucially, Mark's friend waited until he was on his deathbed before revealing his sighting to Mark. Despite knowing his interest in the Mothman phenomenon, he kept it secret right up until the very end.

It is worth pointing out that the dying veteran's report tallies with those of other residents of the Ohio River Valley during that same period, including the sightings that took place in and around Point Pleasant—they perceived the entry as being a giant blackbird, rather than something human shaped. Perceptions morphed with time, becoming less avian and more humanoid with each subsequent retelling.

As Mark continued to tell his story, Richard and Tobias took three steps to the left, seeking out a patch of shade well away from the merciless sun.

"This area has a lot of very ancient energy," Mark continued after taking a moment to collect his thoughts. "There have always been lights in the sky reported around here. My great-grandparents talked about them, believing that the lights were a sign of somebody about to die."

Tobias's friend Ashley interjected with the story of the disappearing airship, a strange tale dating back to October of 1931, when multiple eyewitnesses reported seeing an airship—something akin to a Zeppelin or a blimp—crash into the hills beyond Point Pleasant and Gallipolis. Shortly before 3:00 p.m. in the afternoon of October 10, the flying craft was seen in the skies over Gallipolis, in apparent distress, plunging toward the earth. It tore in half and caught fire prior to impact.

Before the dirigible hit the ground, three occupants jumped out and deployed parachutes. A police-led search party composed of citizen volunteers was deployed, hiking up into the hills where the airship was last seen going down, but no survivors were found. Nor any wreckage or debris. An airship crash is hardly a subtle thing. Smoke, flames, twisted and burned metal would be the norm. Planes made aerial searches of the region. Nothing out of the ordinary was ever discovered.

Investigation revealed that no airships were missing. On October 14, *Bluefield Daily Telegraph* reader C. C. Thornton suggested that a flock of birds may have been mistaken for an airship—a stretch at best, and an explanation that would come back into vogue some thirty-five years later, when it was used by some to explain the Mothman sightings.[73]

---

73. "Blimp Is Believed Down in Wooded Section of State."

Nearly a century later, the mystery of the crashing, disappearing airship has no satisfactory explanation. Could it somehow be tied in with the anomalous lights in the sky, the appearance of Mothman, and other episodes of bizarre activity that have taken place in the vicinity?

"A lot of strange stuff happens on and around this river," Marc nodded in the direction of the Ohio. "There's some darkness around there. Not all those strange things are positive …"

Thanking him for his time, Tobias and Richard headed to a food truck to get something to eat. If he really was the reincarnation of one of the poor souls who died in the collapse of the Silver Bridge—and he certainly believed himself to be—then it was remarkable that Mark felt so comfortable selling his wares just a few feet away from the site of his death in a past life. He seemed to truly have made peace with it, which was all that really mattered.

As the day drew to a close, it was time for Tobias to hit the road for home. Erin, Mike, and Richard stuck around for the second day of the festival. It was still immensely popular, even for a Sunday, but not quite as packed as it had been the day before. As she was browsing for souvenirs, Erin casually fell into conversation with a local resident, one who had an interesting perspective on the Mothman phenomenon. Beckoning Richard over to an outdoor vendor table situated beneath a comfortable shade structure, Erin introduced him to Marqkita.

"Growing up around here, we heard all about the Mothman," Marqkita began. "We were always told never, *ever* to go to the birdhouses, or we'd get our butts spanked for it."

Richard must have appeared puzzled, because she explained that the birdhouses were a local nickname for the derelict buildings located in the TNT area—not the igloos, but rather, the old power station and attendant structures. The bird in question was therefore the Mothman, and the ruined power plant its house.

One day, Marqkita had asked her father whether he thought the Mothman was real. Some locals had dismissed the sightings as being nothing more than some local kids who were high on drugs—a claim for which there is absolutely no evidence. Although teenagers did indeed treat the TNT area as a place for racing cars, recreational drinking, drug use, and hooking up, those who interviewed the first eyewitnesses found them to be both credible and apparently unimpaired.

"What do *you* think?"

"I'm a believer," she replied without hesitation. "I don't know what Mothman is, but it's real. There's a reason why Mothman popped in and popped back out again for a relatively short while, and we don't really see him anymore. It's like … a blip. A blip in reality."

To that day, Marqkita said she still wouldn't go out to the TNT area after dark... though she changed her mind when Richard asked whether she'd do it for fifty bucks. Cash was king, it appeared, and the Mothman couldn't compete with the almighty dollar.

"I wouldn't go off the road, though," she clarified, "and at the first sign of anything out of the ordinary, I'm getting the heck out of there."

Going into one of the igloos was definitely out, then.

Thanking Marqkita for her time, Richard shook her hand and that of her partner, then headed across the street to the Mothman statue. There was a line roughly two hundred people long, waiting to get their photograph taken with the Point Pleasant icon. He shook his head, somewhat bemused at the crowd, which snaked its way along the length of the entire block.

Finally, he made his way back into the throng to find Erin and Mike. It was getting late in the day on Sunday afternoon, and although the crowds were thinning out, there were still plenty of Mothman enthusiasts swarming the streets of Point Pleasant.

As the trio made their way back to the car and hit the road out of town, crossing over the replacement for the Silver Bridge for what may be the last time, Richard was busily dictating notes. He couldn't help but wonder what his coauthor would make of it all.

CONCLUSION

All of this leaves us with one burning question, hanging in the ether like Mothman hovering over a stunned West Virginian: What the hell are people seeing?

Like most aspects of this phenomenon, the answer to this question is complicated and often unclear. There is a nuance necessary to the examination of high strangeness that must allow for multiple answers to the same question, and ultimately, none of them may end up being completely satisfying.

First, we begin with the prosaic explanations. Is this all a hoax? Absolutely not.

Could at least some of it be? Absolutely.

There are many reports that have been circulated that represent unverified information from apparent witnesses who have never been identified. Sometimes, like in the case of many such reports from O'Hare International Airport in Chicago, the details of those reports vary considerably from what has been received from more credible sources. Still other reports might include details that conflict with verifiable information, like the witness reporting the wrong phase of the moon or incorrect weather data for the time and place they claimed for their sighting. This doesn't definitively prove that they're hoaxes, but anyone with an ounce of intellectual honesty must admit that it does make them significantly less credible.

What about misidentifications?

This needs to be taken seriously as a strong possibility for a not insignificant number of sightings, despite its unpopularity among Mothman enthusiasts. Pushback to any misidentification hypothesis is understandable after West Virginia University wildlife biologist Robert L. Smith proposed the winged creature seen in West Virginia and Ohio in the 1960s was nothing more than a sandhill crane that had gotten lost—a controversial-at-best opinion that completely ignored the testimony of eyewitnesses. Similarly, great horned owls and other large native species have been used as explanations for Mothman, with a similarly chilly reception from those interested in the phenomenon.

However, in a baffling irony that one could easily be forgiven for attributing to some Cosmic Joker, there are certain cases with significant evidence in favor of a misidentification, especially among reports out of the Midwest.

Perhaps most notable among these is an incident that occurred in Chicago's Pilsen neighborhood just before 3:00 p.m. on Wednesday, May 9, 2018.

The witness was bicycling to work when he noticed a man and woman standing on the sidewalk looking at something in the sky, the man pointing at whatever it was.

Curious, the witness looked skyward to see what he would later describe as a batlike creature resembling "someone in a wing suit."

Luckily, this witness had a GoPro camera attached to his bike helmet and decided to follow the strange being to record as much footage as he could. He would later very helpfully send this footage, including the original SD card it was saved on, to The Singular Fortean Society. With the witness's help, still images of the creature were isolated from the video and enlarged.

It was a bird.

Likely a very large bird, maybe a heron considering its silhouette, but a bird nonetheless.

There are two species of heron native to the Lake Michigan area that are likely candidates to explain some sightings: the black-crowned night heron and the great blue heron.

Black-crowned night herons have seen something of a resurgence in the Chicago area and could explain many of the sightings seen at night in that city. These nocturnal herons are about two feet in length with a roughly four-foot wingspan, and noticeable crests extend from the back of their head. According to the Lincoln Park Zoo's Urban Wildlife Institute, between 2009 and 2016, the number of paired black-crowned night herons living in Lincoln Park exploded from twenty-four to three hundred, likely due to the destruction of the wetlands they usually call home, which drove them into the city.

Their cousin, the great blue heron, is even larger at around four feet long with a wingspan of up to seven feet; this bird is often described as looking like a prehistoric flying dinosaur. A sizable population of these modern dinosaurs can be readily found at the Chicago Botanic Garden just north of the city.

Herons are migratory birds, most often found in the states bordering Lake Michigan from spring until fall—a particularly popular period for winged creature sightings in these areas. In addition, climate change has affected these birds' migration patterns, and they are coming north earlier and staying longer because of rising temperatures. They can even be found wintering in the Midwest during mild winters—something seen more and more as global temperatures climb.

Herons prefer wetlands and will most often be found near rivers and other large bodies of water. Chicago—and the entire Midwest—has its fair share of those, and many sightings have been near them.

Naturally, herons can't explain every sighting, but they do explain some.

It's possible that this flap of sightings represents two disparate but concurrent phenomena: sightings of herons are more common in areas inhabited by humans due to the destruction of their habitat combined with climate change, and an older series of historical and ongoing sightings of something that could well be paranormal in nature.

This helps explain a strange disparity in the sighting reports. There are essentially two sighting profiles for the Mothman phenomenon. In one, you have sightings, often during the day, of a winged creature seen flying at some distance, with no unusual fear felt by the witnesses or any associated paranormal phenomena; and in the other, you have much closer sightings of a winged humanoid that is often on the ground for at least part of the encounter, with distinctively paranormal attributes like hypnotic glowing red eyes and an unnatural aura of fear, often accompanied by similarly unusual phenomena, including UFOs, poltergeists, and other such anomalies.

That first profile is much likelier to be a heron—or similarly mundane animal (even butterflies in photos have occasionally been mistaken for the Mothman)—while the second is something seemingly far stranger.

But what could be behind such high strangeness? First, we should rule out the unlikeliest possibilities.

Mothman is almost certainly not any mundane undiscovered biological species.

For a breeding population of such a thing to exist, we would have to see its effect on the environment, and yet, there is no evidence of predation in the areas in which it is reported. Not to mention, for a breeding population to exist, it should be inevitable that their habitat be discovered. Any animal seemingly occupying the same location as so many humans should make discovery a foregone conclusion, but here we are, no closer to identifying any of the aspects of such an animal's ecology.

An interesting twist on this is the hypothesis that Mothman is a mutated, but otherwise mundane, member of the local fauna.

The TNT area outside of Point Pleasant was used by the Department of Defense to create dynamite for the war effort from 1942 to 1945. The process of making TNT is highly toxic, and the industrial area, process facilities, and industrial wastewater disposal system were all contaminated. The site was decommissioned and supposedly decontaminated in 1945, after which it was deeded to the state of West Virginia and made into the McClintic State Wildlife Management Area. In 1981, liquid waste produced during the TNT manufacturing process was observed near a pond in the wildlife preserve, and EPA and state investigations revealed that the groundwater and surface water were contaminated with

TNT and its by-products. The site is listed as toxic and is currently being addressed through federal actions.

Despite this, to date, there is no evidence that the area has created any mutants that fit Mothman's description, and the general issues with trying to explain the phenomenon as a mundane biological being remain.

Nor do we have any credible evidence to suggest that Mothman is being brought here aboard an extraterrestrial spacecraft. It being impossible to prove a negative, we can't rule it out entirely, but it's not supported by anything we've seen so far.

That being the case, we might consider Keel's Ultraterrestrials. Such an entity could, in theory, be responsible for any number of claimed paranormal experiences. But the very nature of their being means we don't—and possibly cannot—understand who and what they truly are. That aside, there are a few things we might consider.

Keel proposed that these Ultraterrestrials exist as high-frequency energy, and it was through their mastery of the Superspectrum that they were so easily able to manipulate humankind.

The implied connection to consciousness within Keel's work has continued to gain popularity in the twenty-first century. But what is consciousness? As Tobias wrote in his work *Strange Tales of the Impossible*:

"When most people think of consciousness, they think of one of three commonly accepted dictionary definitions: either the state of being awake and aware of one's surroundings, the awareness or perception of something by a person, or the fact of awareness by the mind of itself and the world. These definitions seem simple enough, describing the basis for understanding and interacting with our environment. And they are simple, right up until we really try to understand them.

"Nobody can seem to agree on what exactly consciousness is or where it comes from. Some, like scientific materialists, think it arises out of the complexities of our brains; a reaction to stimuli that forms an illusion of something more. Still others, such as adherents of biocentrism, posit that consciousness creates the universe, and that without it nothing we observe would exist, at least not in the state in which we observe it. Even more speculation exists within a variety of other disciplines, too, regarding the nature of consciousness and whether it sits exclusively in the human brain, including the conundrum of, if it does exist outside of our bodies, does it do so as a projection from the brain or does it originate from outside of us?"

So, were we to assume that consciousness does exist outside of the human body, then its connection to the otherworldly begins to come into focus.

He would later write in his *Yuletide Guide to High Strangeness*, "I can't shake the feeling that there is some aspect of these phenomena that is enormous and just out of sight, a depth of paranormality concealed beneath the surface of consensus reality. And it touches everything, connecting it—UFOs, ghosts, cryptids, magick, all of it.

"What if that connective force just beneath the surface is consciousness itself, a universal consciousness that our brains merely act as antennae to receive, and upon which our personalities and unique consciousnesses are imprinted like waves in an ocean; individual, yet still part of the whole. Sensory organs for the universe to experience itself.

"Perhaps what we're dealing with is one or more species more native than we are to this hidden reality, a being or beings with an unimaginably expansive reach, wielding power too vast for us to understand, and capable of effortless manipulation of us. Like sharks they traverse this ocean in all dimensions, utterly alien to our understanding, while we are doomed to flounder, barely able to swim.

"Maybe there's something that we call faeries or ghosts or extraterrestrials but is something else entirely; something with which humanity has been interacting for a very long time that has, at different times, gone by many different names, or has at least been interpreted by us in many different ways. In this ecology of consciousness there might be a variety of inhabitants, of whom any may or may not have been misidentified to some extent throughout the ages.

"The truth of this being or beings is completely unknown or, far more unsettlingly, perhaps it is unknowable. Not because it's ultimately impossible for the human mind to comprehend it, but rather, because they don't want us to know."

Unlike Keel's Ultraterrestrial hypothesis, this doesn't depend on a single explanation for every paranormal phenomenon, but rather posits that a series of individual but related phenomena could be to blame. Sort of like how bears and wolves might both live in the same forest, and while they do share many qualities, they are not the same thing. To expand this even further, let us imagine that ghosts really are the surviving consciousnesses of deceased persons. If what we are dealing with are entities that inhabit an environment accessible only by consciousness, then ghosts and aliens might have much more in common than previously thought. They wouldn't be precisely the same, but they would certainly share many qualities, and perhaps that is why the phenomena reported around them bear so many similarities.

All of this is to say, quite honestly, we have absolutely no idea what the actual nature of these phenomena are. Paranormal investigators and researchers are very much the blind leading the blind when it comes to informing the public on the more theoretical aspects of these subjects, and often the best we can do is share our imagined vision of what might exist in the darkness. So, take these musings for what you will, but never limit your imagination to any preset boundaries introduced through speculation. There is always the possibility that some unforeseen explanation lurks just out of sight, waiting to be discovered by someone willing to travel off the map when necessary.

BIBLIOGRAPHY

Barker, Gray. *The Silver Bridge*. Saucerian Books, 2008.

Barker, Gray. *They Knew Too Much About Flying Saucers*. University Books, 1958.

Bell, Karl. "The Legend of the Spring-Heeled Jack." Sussex Folk Tale Centre. Accessed January 8, 2015. https://sussexfolktalecentre.org/wp-content/uploads/Bell-Spring-Heeled-Jack.pdf.

Bender, Albert K. *Flying Saucers and the Three Men*. Saucerian Books, 1962.

Bielawa, Michael J. "Bridgeport's UFO Legacy: Men in Black and the Albert K. Bender Story." Bridgeport Library: Bridgeport History Center. https://bportlibrary.org/hc/authors/bridgeports-ufo-legacy-men-in-black-and-the-albert-k-bender-story/.

"Blimp Is Believed Down in Wooded Section of State." *Bluefield Daily Telegraph,* October 11, 1931.

Castelow, Ellen. "Spring Heeled Jack." Historic UK. Accessed January 8, 2025. https://www.historic-uk.com/CultureUK/Spring-Heeled-Jack/.

"Couples See Man-Sized Bird ... Creature ... Something." *Point Pleasant Register*, November 16, 1966.

"Eight People Say They Saw 'Creature.'" *Charleston Gazette*, November 18, 1966.

"Explosion Reported at Munitions Storage Site." *The Register-Herald*, May 17, 2010. https://www.register-herald.com/news/state_and__region/explosion-reported-at-munitions-storage-site/article_bb3ccb8d-cc8c-549d-9106-ad3a07167917.html.

Fort, Charles. *The Complete Books of Charles Fort*. Dover Publications, 1974.

Hyre, Mary. "Winged, Red-Eyed 'Thing' Chases Point Couples Across Countryside." *The Athens Messenger*, November 16, 1966.

Keel, John A. *The Eighth Tower: On Ultraterrestrials and the Superspectrum*. Anomalist Books, 2013.

Keel, John A. *The Mothman Prophecies*. Tom Doherty Associates, 1991.

Keel, John A. *Operation Trojan Horse*. Anomalist Books, 2013.

Lighty, Chandler. "The Crawfordsville Monster." Hoosier State Chronicles, October 26, 2013. https://blog.newspapers.library.in.gov/crawfordsville-monster/.

Tillman, Nora Taylor. "UFO Sighting? No, Just Google's 'Rogue' Balloon." Space, May 28, 2014. https://www.space.com/26035-google-ufo-was-rogue-balloon.html.

Vaughn, Chad. "Organizers: Over 15,000 Attend Mothman Festival." WV News, September 19, 2023. https://www.wvnews.com/rivercities/gallipolis/news/organizers-over-15-000-attend-mothman-festival/article_2a58215c-54cf-11ee-a69e-73e37bb09a91.html.

Wayland, Tobias. "Flying Humanoid Fever Hits the Windy City, an Interview with Manuel Navarette." A Singular Fortean, June 2017. https://www.singularfortean.com/flying-humanoid-fever-hits-the-windy-city-with-manuel-navarette.

Wayland, Tobias. "Flying Humanoids over Chicago, an Interview with Lon Strickler." A Singular Fortean, July 2017. https://www.singularfortean.com/flying-humanoids-over-chicago-with-lon-strickler.

Wayland, Tobias. *The Lake Michigan Mothman: High Strangeness in the Midwest*. Singular Fortean Publishing, 2019.

Wayland, Tobias. *The Singular Fortean Society's Yuletide Guide to High Strangeness*. Singular Fortean Publishing, 2023.

Wayland, Tobias. *Strange Tales of the Impossible*. Singular Fortean Publishing, 2021.

Wayland, Tobias. "A Timeline of the Lake Michigan Mothman Sightings So Far." A Singular Fortean, June 10, 2017. https://www.singularfortean.com/news/2017/6/10/a-timeline-of-the-chicago-flying-humanoid-sightings-so-far.

Wayland, Tobias. "Witness Reports Harrowing Encounter with Winged Humanoid near Kaneville, Illinois." A Singular Fortean, May 3, 2024, https://www.singularfortean.com/news/2024/5/2/witness-reports-harrowing-encounter-with-winged-humanoid-near-kaneville-illinois.

ACKNOWLEDGMENTS

Few books are created in a vacuum, and *Mothman* is no exception. The authors of this work would like to acknowledge the contributions made to Mothman literature by the plethora of other researchers who have also followed this same strange road, a number of whom paved the way for this and other Mothman books yet to come.

The authors would also like to thank the following people for supporting them or contributing their perspectives and assistance to the project in a multitude of ways.

- *Laura Estep,* who held down the fort while her husband went out looking for Mothman
- *Emily Wayland,* for all her support and inspiration
- *Erin & Mike Taylor,* good companions to have in a dark and eerie TNT complex
- *John EL Tenney* for offering his thoughts on the Mothman phenomenon
- *Mark May* for graciously sharing his experiences
- *Joshua Cutchin,* one of the last true Forteans

- *Richard Hatem*, a consummate writer and entertainer
- *Ashley Hilt*, a great friend and investigator
- *Ken Gerhard*, monster hunter extraordinaire
- *Chad Lewis*, a great researcher and storyteller
- *Jonathan Lane* for welcoming a couple of weirdos into their home
- *Shana and Barb Clippert* for returning to the site of their experience
- The friendly and forthcoming people of Point Pleasant

NOTES